Design of Miniaturized Variable-Capacitance Electrostatic Energy Harvesters

Seyed Hossein Daneshvar • Mehmet Rasit Yuce
Jean-Michel Redouté

Design of Miniaturized Variable-Capacitance Electrostatic Energy Harvesters

 Springer

Seyed Hossein Daneshvar
Monash University
Melbourne
VIC, Australia

Mehmet Rasit Yuce
Monash University
Melbourne
VIC, Australia

Jean-Michel Redouté
University of Liège
Liège, Belgium

ISBN 978-3-030-90254-4 ISBN 978-3-030-90252-0 (eBook)
https://doi.org/10.1007/978-3-030-90252-0

This Springer imprint is published by the registered company Springer Nature Switzerland AG
The registered company address is: Gewerbestrasse 11, 6330 Cham, Switzerland

To the colour and a beautiful meaning of my life, Motahareh, whom I would not be grateful enough dedicating my whole life to.

–S. H. Daneshvar

Preface

Recent achievements in the integration technologies and implantable medical devices and considerable developments in the IoT have made energy harvesting systems an attractive proposition in many innovative applications. As a result of this trend, the importance of having an in-depth understanding of energy harvesting systems is higher than ever before for engineers and researchers promoting these novel areas. Kinetic energy sources are widely available in our environment as well as in our body. This feature makes energy harvesting from kinetic energy sources a promising powering solution in diverse applications such as sensor network, IoT, and medical applications. This book gives the reader an overview of kinetic energy harvesting systems and their applications and a detailed understanding of variable-capacitance electrostatic harvesters (VCEHs). The content is approached from a theoretical as well as a practical point of view, and general challenges in designing kinetic energy harvesting systems are covered in this book.

Energy harvesting methods are investigated as an alternative or prolonging solution to batteries in many emerging applications. The usage of batteries is undesirable or even impractical in these applications due to the fact that batteries are bulky, are heavy, and have a limited lifetime. VCEHs have a superior performance compared to piezoelectric and electromagnetic harvesters in applications with extremely low form factor or when the kinetic energy source operates at lower frequencies. MEMS compatibility and implementation flexibility are other features of this type of harvesters that make them even more desirable. In this book, analysis techniques and optimization solutions, different circuit structures, and practical considerations are discussed in detail.

The first part of the book presents the fundamentals of energy harvesting systems and variable-capacitance electrostatic harvesters (VCEHs). Switched capacitors and asynchronous VCEHs comprise the second part of the book. This part includes structure-specific analysis and optimization solutions for each of these harvesters. The final part develops an experimental setup and practical point of view in implementing electrostatic harvesters. Inductor-based VCEHs are discussed in this part, and a switching technique for miniaturizing this type of VCEHs is presented.

Melbourne, VIC, Australia   Seyed Hossein Daneshvar

Melbourne, VIC, Australia   Mehmet Rasit Yuce

Liège, Belgium   Jean-Michel Redouté
September 2021

Contents

Chapter 1
Kinetic Energy Harvesting Systems Overview

Abstract Batteries were the main power supply for implantable and sensor networks for years. However, new and emerging medical devices and treatments are demanding for more effective, flexible, and long-lasting powering solutions. In the quest to fulfil these new requirements, researchers have been investigating new solutions in making batteries as well as adapting and improving energy harvesting methods. Therefore, standalone energy harvesting systems and hybrid systems where an energy harvesting system is used to prolong the life span of a rechargeable battery are presented in literature. An overview of energy harvesting systems for providing electronics devices with sufficient power is presented in this chapter. The main focus in this chapter is on overview and methods of kinetic energy harvesting systems and their applications.

1.1 Energy Solutions

Providing the electronic devices with required power is one of the main design challenges in many sustainable applications. Different solutions in powering medical devices and sensor networks are reviewed in this section. Depending on the requirements of an application, each of these solutions may become the better choice owing to the fact that each of these solutions has different features, weaknesses, and strengths.

1.1.1 Battery

Batteries are capable of holding high energy densities and they are reliable energy sources. These features are especially of interest in applications with limited access, or where no environmental energy source exists for energy harvesting from, or where energy harvesting solutions are not able to generate the required power density. Examples of these applications may be specific medical or implantable devices. Flexible and biodegradable batteries and super-capacitors are emerging

© The Author(s), under exclusive license to Springer Nature Switzerland AG 2022 1
S. H. Daneshvar et al., *Design of Miniaturized Variable-Capacitance Electrostatic Energy Harvesters*, https://doi.org/10.1007/978-3-030-90252-0_1

candidates for powering implantable electronic devices, and a review on them is presented in [1].

On the other hand, batteries are heavy, bulky, and have a relatively short lifetime. These features make the usage of batteries undesirable or even impractical in many applications. Several surgeries may be necessary for changing the battery of an implantable health-related electronic device such as cardiac pacemakers. In an individual-centred health care system, it is possible to predict or even prevent diseases by continuous monitoring of vital signs. These systems are suggested instead of hospital-centred health care systems and lead to achieve **pervasive** and **personalized** healthcare which are the aims of p-Health. A review on wearable medical systems for p-Health has been presented in [2]. Numerous nodes for monitoring of different symptoms and transferring data are necessary in these systems, and powering all the nodes makes the whole system complex in terms of weight and sustainability. The same issue appears in environmental monitoring systems as the sensor network is huge, and replacing the batteries of each node is a non-trivial task.

1.1.2 Energy Harvesting

In this method, the existing non-electrical energy forms in our surroundings are converted to electrical energy using different types of transducers. The generated electrical energy is then used to supply the required energy of electronic devices. This solution could be implemented in a smaller volume with lighter weights compared to batteries. Although the power density of this method is not as much as batteries in most cases, but recent achievements in electronics proved even nW power level is enough for powering many circuits especially in biomedical applications [3–5]. Therefore, the applications where this method can be solely used have been widened.

Output power of batteries decreases dramatically after few years, while energy harvesting systems are capable of providing power for much longer time. This suggests that repeated surgeries for battery replacement of implantable devices could be avoided with successful implementation of energy harvesting methods. Moreover, battery replacement of numerous sensor nodes in health or environmental monitoring applications could be eliminated.

A wide range of implantable medical devices and emerging treatment solution have been introduce in recent years. This trend with a variety of new requirements demands a revised view of powering electronics devices. Recently, Implantable MicroSystems for Personalized Anti-Cancer Therapy (IMPACT) project has been introduced to develop a new method of cancer treatment [6]. A bunch of sensors are injected to the cancer area which monitor and measure the necessary parameters during the treatment. These sensors identify the position of Radio-Therapy (RT) resistant cancer cells, and the data is used to maximize the damage to these cells. Powering these sensors with independent batteries is impractical, since this will increase the size of sensor nodes. One of the challenges in this research is powering the sensor nodes.

Vagus Nerve Stimulator (VNS) implantation to treat seizures, brain implants to stimulate brain, and monitoring brain activities are other examples in implantable devices requiring power solutions. Sustainability would be an issue if the numerous devices that are connected to Internet of things (IoT) scheme will be powered only by batteries. These applications and many emerging ones are the examples where providing the required energy is a major challenge. Therefore, not only that this challenge appears in well-known areas such as finding solutions for decreasing or eliminating surgeries for battery replacements but also many emerging technologies like new cancer treatments demand for effective solutions for this challenge. Hence, the field of powering electronic devices has attracted many researchers in developing variant of solutions in different fields from new battery generations to finding more efficient energy harvesting methods and energy sources from our surroundings.

1.1.3   Hybrid Solution

The reliability of a powering solution that is only based on energy harvesting techniques is yet an issue. Moreover, the output power of an energy harvesting system is directly dependent on the available instant energy in the energy source. Therefore, using a powering solution that is solely based on energy harvesting methods in applications with high reliability requirement or where the target device requires instantaneous high level of energy is not practical. To get the benefits of both above-mentioned powering solutions for these applications, hybrid energy harvesting systems are introduced in which the generated energy is stored in a rechargeable battery or a super-capacitor. The hybrid system has a longer lifetime compared to the battery based systems and is more reliable than the energy harvesting based systems.

1.2   Overview of Energy Harvesting Systems

Components of an energy harvesting system are shown in Fig. 1.1. In this figure, the electromechanical coupling refers to the necessary mechanical part that couples the transducer to the energy source, and the load is the target electronic device demanding electrical power. Researchers have conducted numerous studies in each of the components of this figure. In the following subsections a brief review on the reported energy sources, transducers, and output circuits is presented.

Fig. 1.1 Components of a typical energy harvesting system

1.2.1   Energy Source

A variety of energy sources that have been used in powering nodes of sensor networks, wearable and implantable medical devices are reviewed in this section.

1.2.1.1   Environmental Sources

The environment is a rich source of energy in diverse forms, and many of these sources have been used in energy harvesting systems so far. Solar cells generate electrical energy from the sun light. In [7], two solar panels are placed over a patient's shoulders to harvest solar energy, whenever the person is exposed to sun radiation. This type of energy harvesting system can provide relatively large amount of power; however, they do not generate power when there is no light. Temperature gradient has also been used to obtain energy from human body. This type of energy harvesting system employs a thermoelectric generator (TEG). The TEG is placed where there is a temperature difference between its two sides. It is estimated that this generator is capable of producing $60\,\mu W/cm^2$ power from the temperature gradient between the body and the environment [7]. In this case, the TEG is typically placed close to the skin such that one side of the TEG faces the skin and the other side faces inside the body. The size of the TEG used in such systems is an issue and their efficiency heavily depends on environmental conditions.

1.2.1.2   Power and Data Links

In this method, an emitter should transmit electromagnetic waves and a receiver that is integrated to the electronic device collects this wave as transmitted power. This link can also be utilized as data link, in case transferring data between a source and an implantable device is essential. Sensor nodes are an example of these applications. RF waves and inductive link are the solutions that have been presented in this category.

Radio Frequency (RF) waves are the source of energy that is found almost everywhere nowadays. Electronic devices receive RF waves from an emitter, which might be far away from the harvesting system. This solution is especially used in powering numerous sensor nodes where battery replacement is not practical [8]. A circuit capable of converting the RF waves to DC power should be integrated in the target device. If the power of the received signal is below $100\,\mu W$, the applied voltage to the RF-DC converter is only 0.3 V [8]. Therefore, utilizing a rechargeable battery and power efficient design of the converter are critical for reliable operation of target devices. An energy harvesting system based on this form of energy is also proposed in [9]. The energy that can be harvested decreases dramatically with respect to the distance from the RF source.

Due to the high absorption coefficient of RF waves in the body tissue, this type of power link is not feasible for powering implantable devices. Inductive power and data link are used for this purpose instead. In inductive link solution, an emitter coil and a receiver coil are used to make a power link between the source of energy and the electronic device. This solution is used in powering implantable devices [10]. The efficiency and the amount of power that can be transferred between the source and the electronic device depend on the coils' size. This way, the harvester size limits the efficiency and the amount of deliverable power. Therefore, the main challenges in this method are optimizing the amount of transmitted power and the amount of data [10, 11]. Another challenge of designing such systems is controlling the transmitted power. Variation of the load and the mutual coefficient might result in increasing the temperature of the electronic device which in turn makes irreversible consequences on the body. Moreover, the voltage or the current of the electronic device may decrease as a result of insufficient transmitted power. Therefore investigation of solutions for compensating the load variation or the mutual coefficient is desirable. To perform this compensation the data about the used power in the receiver coil is sent to the source coil, and this data defines the amount of power that should be transmitted [12].

1.2.1.3 Kinetic Energy Sources

This energy exists in moving objects. The kinetic energy sources that have been used for energy harvesting purposes can be categorized to two main groups: vibration and human body movements.

A variety of vibration energy sources, their fundamental frequency, and amplitude are presented in [13]. Office windows next to busy roads, microwave oven, and blender casing are few examples of reported vibration sources. The possibility of powering wireless sensor nodes with the energy harvested from vibration is investigated in [14]. The main challenge in energy harvesting from vibration is matching the resonance frequency of the harvester with the energy source frequency. The size of a vibration based harvester needs to be larger for lower frequencies to have matched resonance frequency. Therefore, using this type of kinetic harvester for lower frequency movements is not optimal in terms of size.

Despite vibration, the movements of human body are normally of low frequency. The design of the harvesters for energy harvesting from these low frequency movements is generally different from vibration based harvesters. In [15], an electromagnetic micro-generator, with a volume of $1.5\,\text{cm}^3$, is used to extract $3.9\,\mu\text{W}$ power from the movement of the ankle. A micro-scale piezoelectric generator is used to harvest energy from the blood pressure fluctuation of human body [16]. An electromagnetic generator with a stator of ceramic magnets and a sliding plastic coil as the rotor is proposed in [17, 18]. This generator follows the up and down movements of human shoulders and is capable of generating 90–360 mW depending on the walking pattern. The acceleration and the frequency of different locations on human body while running or walking are measured in [19], to find the maximal available power at each of the locations. New piezoelectric materials are investigated for energy harvesting from heart, lung, and diaphragm [20].

1.2.2   Transducer

Typical micro-scale harvesters that have been used for energy harvesting from kinetic energy sources are piezoelectric, electromagnetic, and electrostatic. In this section, a brief review on these transducers for powering implantable and health-care related electronic devices is presented.

1.2.2.1   Piezoelectric

Piezoelectric micro-scale generators are inexpensive, compact, and have no moving part, and they can be implemented simply. Several structures of this kind of micro-scale generators are reviewed in [21]. However, a high pressure and low amplitude energy source is needed for considerable power in the output of this type of micro-scale generator. In [13] and [14], piezoelectric micro-scale generators have been used in energy harvesting from vibrations. The first prototypes of piezoelectric harvesters for energy harvesting from human body while walking are introduced in [22–24]. These harvesters use the bending of a shoe insole during walking to generate electrical energy. A hexagon-shaped sneaker insole which is designed for energy harvesting from human walking generates 1.3 mW with a movement frequency of 0.9 Hz. Moreover, energy harvesting from heart, lung, and diaphragm based on new piezoelectric materials is presented in [20]. Piezoelectric harvesters produce a relatively high voltage with a low current output signal which is not directly usable for electronic devices fabricated with new technologies [21]. Therefore, a conditioning circuit is required to make their output appropriate for powering new generation of electronic devices. A number of circuits which can be utilized for this purpose are discussed in [25].

1.2.2.2   Electromagnetic

Electromagnetic generators have low voltage and high current output, which is more proper for today's technology of fabricating electronic devices. Several researches employing this type of micro-scale generators are mentioned in Sect. 1.2.1.3. Different structures of this type of micro-scale harvester are resonance, rotational, and hybrid [21]. However, electromagnetic generators used in energy harvesting applications are expensive, and the main shaft of the electromagnetic generator should run at a high enough speed otherwise the output voltage is very low [21]. In cases that the output signals of these generators are not suitable for supplying an electronic circuit directly, an output circuit is used [25].

1.2.2.3 Electrostatic

Electrostatic generators were proposed many years ago [26, 27]. They were not used widely for generating electricity in macro-scale as the amount of energy they can generate is lower compared to electromagnetic generators. However, this type of harvester becomes an attractive candidate in micro-scale applications. The reasons for this are explained in Sect. 1.4.3 by detailing the features of electrostatic harvesters and comparing them with other kind of harvesters in micro-scale applications. Main components of an electrostatic harvester are briefly discussed by an example of this type of harvester in the following section.

1.3 Electrostatic Harvester: An Example

An electrostatic harvester comprises two main parts: a variable capacitor and a switching circuit making the variable capacitor go through different phases of energy harvesting. Figure 1.2 shows the electrostatic harvester that is presented in [28]. The variable capacitor in Fig. 1.2a is designed for energy harvesting from low frequency long amplitude movements. Therefore, the structure of this variable capacitor is of interest in energy harvesting from body movements. The circuit that is used with this variable capacitor in [28] is shown in Fig. 1.2b. The rod rolls from one side to the other side over the printed plates on a glass substrate in Fig. 1.2a. The substrate is coated with a dielectric material. The rod insulates from the printed plates, and the variable capacitor is formed between the rod and the printed plates on the glass substrate. At any time that the rod is aligned with a printed plate area, the capacitance of the variable capacitor is maximal. At this moment, the rod connects to a battery through the input contact, and the battery charges the variable capacitor. Next, the rod eventually aligns with the area with no printed plate while there is no contact between the rod and the battery. The voltage across the variable capacitor and subsequently the stored energy in it increase. The stored energy in the variable capacitor gets to its maximal value when the rod is fully aligned with the area with no plate. At this moment, the rod connects to the output through the output contacts, and the stored energy in the variable capacitor is transferred to the device connected to the output. The rod continues rolling and eventually become aligned with the printed plate area. A new energy harvesting cycle starts at this moment. During a full energy harvesting cycle, no energy is returned to the battery. This way, the

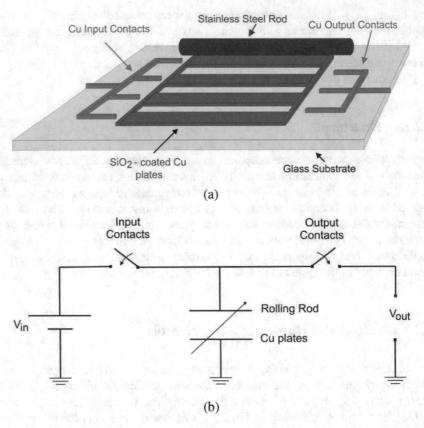

Fig. 1.2 The main components of an electrostatic harvester (**a**) An example variable capacitor (**b**) Non-sustainable circuit harvester

battery eventually depletes, and therefore, this circuit is non-sustainable. Analysis of this non-sustainable circuit is presented in Sect. 2.4.3.1.

As can be seen, the capacitance of the variable capacitor changes several times (depending on the number of printed plates) when the rod moves from one side to other side. This is equivalent to increasing the frequency of a linear movement.

1.4   Scope of This Book

This book discusses electrostatic harvesters in detail as a considerable candidate for energy harvesting from kinetic energy sources. Miniature size applications and in particular powering sensor nodes, wearable and implantable medical devices are targeted throughout this book. This section clarifies why this scope has been selected in this book.

1.4.1 Why Energy Harvesting

Different powering solutions are discussed in Sect. 1.1. As explained, including energy harvesting methods in powering solutions not only is advantageous in a wide range of applications, but also it is the only practical solution in many emerging applications.

1.4.2 Why Kinetic Energy Source

Repetitive surgeries are required for replacing the battery of an implantable device, in case it is powered by a battery. Alternatively, inductive power links may be used for powering implantable devices to remove the need for replacing the battery. However, this method needs the patient to regularly use an external power transmitter to charge an integrated rechargeable battery in the implantable device. Implementing an energy harvesting solution that uses continuous movements of internal organs would eliminate the need for replacing the battery and the need of using an external device regularly. Moreover, the characteristics of these kinetic energy sources are more predictable and less dependent on the environmental conditions compared to energy sources such as temperature gradient and solar.

1.4.3 Why Electrostatic Harvesters

The following features of electrostatic harvesters make them an attractive candidate in a variety of miniature size applications.

1.4.3.1 Micro-scale Implementation Challenges

In medical and micro-scale applications, the available space is limited. If the sizes of a generator scale down with a coefficient of S, electrostatic force scales down with a coefficient between S^1 and S^2, while electromagnetic force scales down with a coefficient between S^3 and S^4 [29]. This is due to the fact that in micro-scales, the quality of the permanent magnets used in electromagnetic harvesters decreases dramatically and the number of turns of the magnetic coils has to be reduced to fit in the limited available space [21]. The main component in an electrostatic harvester is a variable capacitor. The maximal capacitance of this capacitor increases when the distance between its two plates decreases. Therefore, electrostatic harvesters start performing better at some point, compared to electromagnetic harvesters when the size of the system scales down.

1.4.3.2 MEMS Compatibility

MEMS compatible micro-scale harvesters are of interest in System on a Chip (SoC) applications, since they are cheaper and they can be integrated with essential integrated circuits. "Electrostatic transducers are often used as actuators in MEMS devices and thus variable capacitance structures are readily integratable using standard microfabrication technology" [30]. However, MEMS electromagnetic generators are rarely reported [31, 32], since planar magnets have poor properties and the number of coil turns is limited [21]. As a result, the fabrication process of micro-scale electromagnetic harvesters is expensive and complicated.

1.4.3.3 Frequency Dependency

The generated power of a rotary electromagnetic harvester reduces exponentially, if the frequency of the kinetic energy source decreases. This occurs due to winding losses and internal inductance in the structure of electromagnetic harvesters [33]. Figure 1.3 shows the lab test of a rotary electromagnetic generator (Seiko kinetic series) in [33]. The output power of a linear micro-scale electromagnetic harvester [34] is shown to have the same relation with respect to the energy source acceleration in [33]. However, the output power of a micro-scale electrostatic harvester depends linearly on frequency. Moreover, the output voltage of micro-scale electromagnetic harvesters depends on the energy source frequency linearly. However, the output voltage of electrostatic harvesters does not depend on the

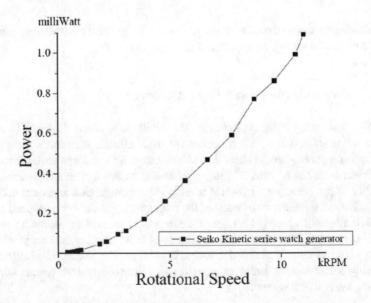

Fig. 1.3 Lab test of Seiko's kinetic generators [33]

energy source frequency and could be adjusted regardless of the energy source frequency.

1.4.3.4 Emerging Technologies

From a practical point of view, emerging technologies, e.g. microfluidics, can be used to fabricate variable capacitors with higher performance [35]. These variable capacitors optimize the output power of electrostatic harvesters even further. A variety of shapes, structures, and mechanisms are proposed in fabricating suitable variable capacitor for different applications. This makes electrostatic harvesters a flexible choice in a wide range of applications. A variable capacitor that is suitable for energy harvesting from long amplitude, low frequency movements is shown in Fig. 1.2a. Another variable capacitor with this feature [35] is explained in more details in Sect. 3.5.1. Using this variable capacitor, two energy harvesting solutions are presented in Sect. 3.5.

1.5 Structure

Chapter 2 covers fundamentals of electrostatic harvesters operation and categorizes this type of harvester. Basic steps for analysing electrostatic harvesters are explained and these techniques are used in a few examples. Chapter 3 discusses switched-capacitor electrostatic harvesters. This category of electrostatic harvesters is optimal in applications with limited available volume, since using bulky inductors is avoided. Two real-world applications using these harvesters are explained in this chapter. The switched-capacitor harvesters that are discussed in Chap. 3 are synchronous. Therefore, the switching events in their circuit should be synchronized with the moments that the capacitance of the variable capacitor is minimal and/or maximal. In Chap. 4, asynchronous electrostatic harvesters are discussed. The switching events in the circuit of these harvesters do not need to be synchronized with the variable capacitor. This simplifies the implementation of the harvester, since the required control circuit for these harvesters is less complex compared to the harvesters in Chap. 3. The conduction losses are reduced in electrostatic harvesters that use an inductor in their circuit. However, using inductor is not optimal in miniature size applications. Steps for analysing electrostatic harvesters that use inductor are detailed in Chap. 5. An improved electrostatic harvester that uses a miniature size inductor is presented in this chapter.

References

1. A. Kim, M. Ochoa, R. Rahimi, B. Ziaie, New and emerging energy sources for implantable wireless microdevices. IEEE Access **3**, 89–98 (2015)
2. X. Teng, Y. Zhang, C.C.Y. Poon, P. Bonato, Wearable medical systems for p-health. IEEE Rev. Biomed. Eng. **1**, 62–74 (2008)
3. Z. Zhu, Y. Liang, A 0.6-V 38-nW 9.4-ENOB 20-kS/s SAR ADC in 0.18-μmCMOS for medical implant devices. IEEE Trans. Circuits Syst. I: Regul. Pap. **62**(9), 2167–2176 (2015)
4. A.F. Yeknami, A. Alvandpour, A 0.5-V 250-nW 65-dB SNDR passive $\Delta\Sigma$ modulator for medical implant devices, in *2013 IEEE International Symposium on Circuits and Systems (ISCAS2013)*, Beijing (2013), pp. 2010–2013
5. H. Tang, Z.C. Sun, K.W.R. Chew, L. Siek, A 5.8 nW 9.1-ENOB 1-kS/s local asynchronous successive approximation register ADC for implantable medical device. IEEE Trans. Very Large Scale Integr. (VLSI) Syst. **22**(10), 2221–2225 (2014)
6. Implantable MicroSystems for Personalised Anti-Cancer Therapy (IMPACT) project being conducted by the University of Edinburgh, Current, Available online at https://impact.eng.ed. ac.uk
7. J.A. Paradiso, T. Starner, Energy scavenging for mobile and wireless electronics. IEEE Pervasive Comput. **4**(1), 18–27 (2005)
8. T. Le, K. Mayaram, T. Fiez, Efficient far-field radio frequency energy harvesting for passively powered sensor networks. IEEE J. Solid-State Circuits **43**(5), 1287–1302 (2008)
9. B. Li, X. Shao, N. Shahshahan, N. Goldsman, T. Salter, G.M. Metze, An antenna co-design dual band RF energy harvester. IEEE Trans. Circuits Syst. I: Regul. Pap. **60**(12), 3256–3266 (2013)
10. A.M. Sodagar, K. Najafi, K.D. Wise, M. Ghovanloo, Fully-integrated CMOS power regulator for telemetry-powered implantable biomedical microsystems, in *IEEE Custom Integrated Circuits Conference 2006*, San Jose (2006), pp. 659–662
11. M. Ghovanloo, K. Najafi, Fully integrated wideband high-current rectifiers for inductively powered devices. IEEE J. Solid-State Circuits **39**(11), 1976–1984 (2004)
12. G. Wang, W. Liu, M. Sivaprakasam, G.A. Kendir, Design and analysis of an adaptive transcutaneous power telemetry for biomedical implants. IEEE Trans. Circuits Syst. I: Regul. Pap. **52**(10), 2109–2117 (2005)
13. S. Roundy, P.K. Wright, J. Rabaey, A study of low level vibrations as a power source for wireless sensor nodes. Comput. Commun. **26**(11), 1131–1144 (2003)
14. S. Roundy et al., Improving power output for vibration-based energy scavengers. IEEE Pervasive Comput. **4**(1), 28–36 (2005)
15. E. Romero, R.O. Warrington, M.R. Neuman, Body motion for powering biomedical devices, in *2009 Annual International Conference of the IEEE Engineering in Medicine and Biology Society*, Minneapolis (2009), pp. 2752–2755
16. M.J. Ramsay, W.W. Clark, Piezoelectric energy harvesting for bio-MEMS applications, in *Proceedings of SPIE 2001*, vol. 4332 (2001)
17. P. Niu, P. Chapman, Design and performance of linear biomechanical energy conversion devices, in *2006 37th IEEE Power Electronics Specialists Conference*, Jeju (2006), pp. 1–6
18. P. Niu, P. Chapman, L. DiBerardino, E. Hsiao-Wecksler, Design and optimization of a biomechanical energy harvesting device, in *2008 IEEE Power Electronics Specialists Conference*, Rhodes (2008), pp. 4062–4069
19. E. Romero-Ramirez, Energy harvesting from body motion using rotational micro-generation. Dissertation, Michigan Technological University (2010). Available online at: http://digitalcommons.mtu.edu/etds/404
20. C. Dagdeviren, B. Duk Yang, Y. Su, P.L. Tran, P. Joe, E. Anderson, J. Xia, V. Doraiswamy, B. Dehdashti, X. Feng, B. Lu, R. Poston, Z. Khalpey, R. Ghaffari, Y. Huang, M.J. Slepian, J.A. Rogers, Conformal piezoelectric energy harvesting and storage from motions of the heart, lung, and diaphragm. PNAS 2014 **111**(5), 1927–1932

21. A. Khaligh, P. Zeng, C. Zheng, Kinetic energy harvesting using piezoelectric and electromagnetic technologies-state of the art. IEEE Trans. Ind. Electron. **57**(3), 850–860 (2010)
22. J. Kymissis, C. Kendall, J. Paradiso, N. Gershenfeld, Parasitic power harvesting in shoes, in *Digest of Papers. Second International Symposium on Wearable Computers (Cat. No.98EX215)*, Pittsburgh (1998), pp. 132–139
23. N.S. Shenck, A demonstration of useful electric energy generation from piezoceramics in a shoe. M. S. thesis, MIT, Cambridge, MA, 1999
24. N.S. Shenck, J.A. Paradiso, Energy scavenging with shoe-mounted piezoelectrics. IEEE Micro **21**(3), 30–42 (2001)
25. G.D. Szarka, B.H. Stark, S.G. Burrow, Review of power conditioning for kinetic energy harvesting systems. IEEE Trans. Power Electron. **27**(2), 803–815 (2012)
26. E.B. Kurtz, M.J. Larsen, An electrostatic audio generator. Electr. Eng. **54**(9), 950–955 (1935)
27. F.H. Merrill, The van de Graaff electrostatic generator. Stud. Q. J. **9**(35), 124–127 (1939)
28. M.E. Kiziroglou, C. He, E.M. Yeatman, Rolling rod electrostatic microgenerator. IEEE Trans. Ind. Electron. **56**(4), 1101–1108 (2009)
29. W.S.N. Trimmer, Microrobots and micromechanical systems. Sens. Actuat. **19**(3), 267–287 (1989)
30. P.D. Mitcheson, T. Sterken, C. He, M. Kiziroglou, E.M. Yeatman, R. Puers, Electrostatic microgenerators. Meas. Control **41**(4), 114–119 (2008)
31. W.J. Li, G.M.H. Chan, N.N.H. Ching, P.H.W. Leong, H.Y. Wong, Dynamical modeling and simulation of a laser-micromachined vibration-based micro power generator. Int. J. Nonlinear Sci. Simul. **1**, 345353 (2000)
32. C.B. Williams, S. Shearwood, M.A. Harradine, P.H. Mellor, T.S. Birch, R.B. Yates, Development of an electromagnetic micro-generator. IEE Proc. Circuits Dev. Syst. **148**(6), 337342 (2001)
33. J. Boland, Micro electret power generators. PhD. Thesis, California Institute of Technology, 2005. Available online at: https://thesis.library.caltech.edu/5228/1/JustinBoland.pdf
34. S.P. Beeby, M.J. Tudor, E. Koukharenko, N.M. White, T. ÓDonnell, C. Saha, S. Kulkarni, S. Roy, Micromachined silicon generator for harvesting power from vibrations, in *Presented at the Fourth International Workshop on Micro and Nanotechnology for Power Generation and Energy Conversion Applications*, Kyoto (2004)
35. T.N. Krupenkin, Method and Apparatus for Energy Harvesting Using Microfluidics, U. S. Patent 8,053,914 B1, 8 Nov 2011

Chapter 2
Electrostatic Harvesters Overview and Applications

Abstract An electrostatic harvester operates with the frequency of an energy source. During each period of operation, a sequence of events (phases of operation) occur, so that an amount of energy is converted into electrical form. These phases of operation are explained in this chapter. Then, the electrostatic harvesters are categorized into sustainable and non-sustainable types according to a system-level point of view. QV diagram is a useful tool in evaluating different aspects of electrostatic harvesters. This diagram shows the relation between the voltage (V) across the variable capacitor and the charge (Q) in this capacitor during different phases of operation. Two naming conventions are suggested in this chapter. Using these naming conventions, the electrostatic harvesters are categorized based on the QV diagram of the variable capacitor and circuit features. Later, calculating the deliverable energy, the net generated energy, and the conduction losses is discussed for electrostatic harvesters. Finally, a discrete analysis method is explained for finding closed-form expressions. The closed-form expressions give an essential insight in the operation of electrostatic harvesters.

2.1 Fundamentals of Electrostatic Harvesters

The fundamental knowledge of electrostatic harvesters is discussed in this section. General phases of operation and system-level structures are explained first. Next, the QV diagram of the variable capacitor in these harvesters is discussed. Later, the electrostatic harvesters are categorized based on the QV diagram and the type of switching events during its phases of operation.

2.1.1 Operation of an Electrostatic Harvester

During a full operating cycle of an electrostatic harvester, the energy source changes the capacitance of the variable capacitor from maximal to minimal and back to maximal again. Assuming each full operating cycle takes T_V to occur, the frequency of changes in the capacitance of the variable capacitor is defined as follows:

$$f_V = \frac{1}{T_V}.$$

(2.1)

In an electrostatic harvester, the following four phases occur in each full operating cycle. The comprehensive operation of an electrostatic harvester may be explained, considering that a full operating cycle including these phases happens with a frequency of f_V.

Phases of Operation

Investment phase: A storage component, e.g. a battery or a reservoir capacitor, charges a variable capacitor (C_V) to an initial voltage, V_r.

Harvesting phase: An energy source changes the capacitance of C_V from C_{max} to C_{min}. As a result of this change in C_V, an amount of energy is converted into electrical form and is stored in C_V. This energy conversion occurs, provided that C_V was initially charged before the start of this phase.

Reimbursement phase: An amount of the converted energy that was stored in C_V during the harvesting phase is transferred from C_V:

- To the load, in a non-sustainable system
- Back to the storage component (the battery or the reservoir capacitor), in a sustainable system

The structure of these systems is explained in the following subsection.

Recovery phase: The energy source changes the capacitance of C_V from C_{min} to C_{max}. Therefore, C_V becomes ready for the next cycle.

2.1.2   System-Level Structure of Electrostatic Harvesters

Phases of operation are explained based on the changes in the capacitance of the variable capacitor in the above. Other essential components to implement a practical electrostatic harvester are discussed here from a system-level point of view. These harvesters are categorized into the following two types in this regard.

2.1.2.1   Non-Sustainable Electrostatic Harvesters

Figure 2.1 shows the general structure of non-sustainable electrostatic harvesters. The battery initially charges the variable capacitor through the investment circuitry. The energy source changes the capacitance of the variable capacitor from maximal to minimal value. This generates an amount of energy in the variable capacitor.

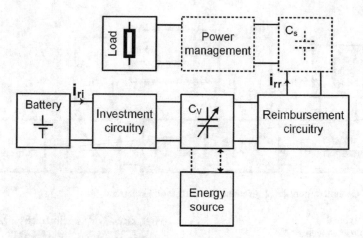

Fig. 2.1 General system-level non-sustainable electrostatic harvesters

The generated energy may be accumulated in a temporary storage component (C_s) through the reimbursement circuitry. The power management block then manages the smooth energy transfer between C_s and the load. However, the existence of C_s and the power management block in this system are not essential, and they may be eliminated in many applications. The generated energy in the variable capacitor may be transferred through the reimbursement circuitry directly to the load, in this system. Therefore, C_s and the power management block are shown with dashed lines instead of solid lines in this system. This cycle repeats after that the energy source changes the capacitance of the variable capacitor from minimal to maximal.

The generated energy in the variable capacitor does not reimburse to the battery that initially charged the variable capacitor. Although the load receives more energy than the obtained energy from the battery (due to the energy generation in the variable capacitor), however the battery depletes eventually. Therefore, this system is not sustainable.

2.1.2.2 Sustainable Electrostatic Harvesters

Figure 2.2 shows the general structure of sustainable electrostatic harvesters. In this system, the storage component (which is depicted as C_r) charges the variable capacitor initially through the investment circuitry. An amount of energy is generated in the variable capacitor. The generated energy is transferred back to the storage component through the reimbursement circuitry. This way, the system becomes sustainable if the energy that is transferred back to the storage component is more than the energy that is obtained from it initially to charge the variable capacitor. The load is connected to the storage component (a large capacitor or a rechargeable battery) through an optional power management block.

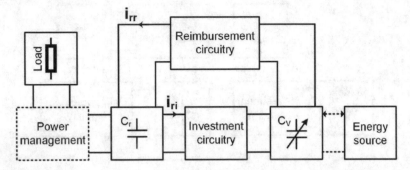

Fig. 2.2 General system-level sustainable electrostatic harvesters

Fig. 2.3 Typical
non-sustainable electrostatic
harvester

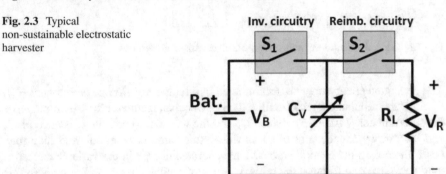

2.1.3 Circuit-Level Electrostatic Harvester Example

Figure 2.3 shows a typical non-sustainable electrostatic harvester. In this figure, the storage component is a battery and the load is resistive. The investment and reimbursement circuitries are highlighted, and each of these circuitries is implemented with a single switch. The voltages across the variable capacitor and the load are shown in Fig. 2.4, along with the status of the variable capacitor. According to this figure and the phases of operation, as explained in Sect. 2.1.1, the operation of this harvester is explained as follows.

In Fig. 2.4, the signals for this circuit are shown for three full operating cycles. The phases are noted for the second cycle; however, the same labels are true for all operating cycles of this harvester. The variable capacitor is charged to V_B at the beginning of the investment phase: S_1 : on, S_2 : off. This happens at the moment that $C_V = C_{max}$. Both S_1 and S_2 turn off, and the variable capacitor is isolated from the rest of the circuit. The energy source changes the capacitance of the variable capacitor from maximal towards minimal, during the harvesting phase. The voltage across the variable increases to nV_B (where n is C_{max}/C_{min}) from V_B during this phase, since the charge is constant in C_V and according to $Q = CV$. This procedure implements the charge-constraint switching scheme that is explained in more detail in Sect. 2.2.1. In the reimbursement phase, C_V connects to the load when $C_V =$

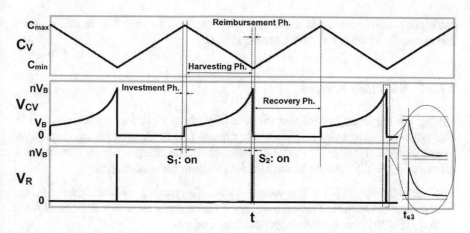

Fig. 2.4 The changes in the capacitance of the variable capacitor and the voltages across C_V and R_L in the circuit of Fig. 2.3

C_{min}: S_1 : off and S_2 : on. This phase is magnified in Fig. 2.4. The voltage across C_V and the load during this phase is

$$V_{CV}(t) = V_R(t) = nV_B e^{-\frac{t}{\tau}}, \qquad t \geq t_{s3}, \tag{2.2}$$

where:

$$\tau = C_{min} R_L. \tag{2.3}$$

Therefore, the whole charge in C_V transfers to the load as long as S_2 is on approximately for more than 5τ (where τ is defined in the above expression). The energy source then changes the capacitance of the variable capacitor from minimal towards maximal during the recovery phase, while S_1 and S_2 are both off. The variable capacitor becomes ready for the next operating cycle at the end of this phase when C_V reaches to its minimal value, C_{min}. The same sequence of events repeats in the next operating cycles.

2.1.4   QV Diagram

In an electrostatic harvester, variations in the voltage (V) across and in the charge (Q) in a variable capacitor during a full energy conversion cycle make an enclosed area in the QV diagram. This area represents a part of the harvested energy in the variable capacitor that is deliverable to the energy storage component. This area is positive if arrows on the path that makes it are in a clockwise direction. The calculation of this energy is discussed in more detail in Sect. 2.2. In this section, fundamentals of the QV diagram are explained.

2.1.4.1 Switching Schemes

During the harvesting phase and the recovery phase (that the capacitance of C_V changes), the voltage (V) across and the charge (Q) in C_V may change according to one or a combination of the below scenarios. The charge-voltage formula of a capacitor ($Q = CV$) should be considered to follow these scenarios:

- Charge-Constraint (CC): The voltage across C_V changes, while the charge in this capacitor is constant. This case occurs if the variable capacitor is isolated from the rest of the circuit when its capacitance changes.
- Voltage-Constraint (VC): The charge in C_V changes, while the voltage across this capacitor is constant. This case occurs if the variable capacitor is connected to battery or a large capacitor (compared to C_V), when its capacitance changes.
- No-Constraint (NC): Both the voltage across and the charge iC_V change. This case occurs if C_V is connected to a capacitor (not large compared to C_V), while its capacitance changes.

2.1.4.2 Possible Paths in QV Diagram

Figures 2.5 and 2.6 show all the possible paths for the QV diagram of the variable capacitor during the harvesting phase and the recovery phase, respectively. The QV diagram of C_V is always plotted in the region that $v, q > 0$. The slope of any line between the points (0,0) and (v,q) in these QV diagrams is equal to the capacitance of C_V. This is evident from the formula $Q = CV$. Dashed lines in Figs. 2.5 and 2.6 show the equivalent lines for $C_V = C_{max}$ and $C_V = C_{min}$. All the paths presenting the harvesting phase have a starting point, **a**, on the line of $C_V = C_{max}$ and a destination point, **b** or **b'**, on the line of $C_V = C_{min}$. On the other hand, the starting point and the destination point are on the line of $C_V = C_{min}$ and $C_V = C_{max}$, respectively, for all the paths presenting the recovery phase.

In Fig. 2.5a, the horizontal line represents a charge-constraint (CC) change in the capacitance of the variable capacitor. This is the case when C_V is isolated from the rest of the circuit during the harvesting phase. The vertical line in Fig. 2.5b represents a voltage-constraint (VC) change in the capacitance of C_V. This happens when C_V is connected to a battery or a large capacitor ($C \gg C_{max}$) during the harvesting phase. If C_V is connected to a capacitor that is not large ($C \ngg C_{max}$) during this phase, the QV diagram follows the path **ab'** instead of **ab**. This line with a negative slope represents a no-constraint (NC) change in the capacitance of the variable capacitor.

Figure 2.5a and b shows the cases that the capacitance of C_V changes under one scheme (CC, or VC, or NC) for the whole duration of the harvesting phase.

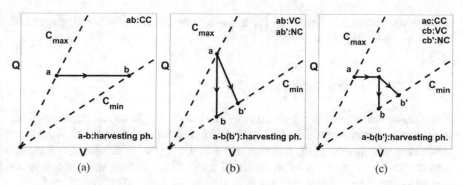

Fig. 2.5 The possible paths in the QV diagram during the harvesting phase when the capacitance of the variable capacitor changes from C_{max} to C_{min}

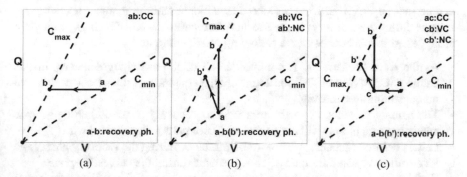

Fig. 2.6 The possible paths in the QV diagram during the recovery phase when the capacitance of the variable capacitor changes from C_{min} to C_{max}

To get to point **b** (or **b'**) from point **a**, a combination of these schemes may also be used. Figure 2.5c shows the case where a combination of CC-VC or CC-NC happens during the harvesting phase. The capacitance of C_V does not necessarily change under one scheme during one phase: it may change under a combination of schemes. Figure 2.6a–c shows the similar cases for the QV diagram of C_V during the recovery phase.

2.1.5 Categorizing Electrostatic Harvesters

Switching electrostatic harvesters are conventionally categorized into charge-constraint and voltage-constraint. This naming is based on that the capacitance of C_V changes whether under charge-constraint or voltage-constraint during the harvesting phase. Although categorizing electrostatic harvesters into these two types covers an important aspect of them, it ignores other features of the electrostatic harvesters. Therefore, the following two conventions are suggested and used throughout this book.

2.1.5.1   QV Diagram-Based Categories

This naming convention is suggested to be used when the focus is on the behaviour of electrostatic harvesters in the QV diagram. In categorizing electrostatic harvesters into either charge-constraint or voltage-constraint types, it is assumed that the capacitance of C_V changes under either of the schemes for the whole duration of the harvesting phase. This assumption is simply not correct for all different types of electrostatic harvesters. Moreover, the switching scheme in the recovery phase affects the QV diagram and ultimately the amount of generated energy in C_V, as can be seen in Fig. 2.6a. However, the conventional naming ignores the switching scheme during the recovery phase. The naming convention based on the QV diagram is suggested in this book with the following considerations:

- As the word "constraint" is common in all of the switching schemes, in this naming "C," "V," and "N" represent charge-constraint, voltage-constraint, and no-constraint, respectively.
- The naming includes two parts representing recovery phase switching scheme and harvesting phase switching scheme. These parts are separated with ":".
- Letters before ":" stand for the switching scheme during the recovery phase, and letters after ":" stand for the switching scheme during the harvesting phase.
- Each of the above parts may include one or two letters depending on whether only one switching scheme or two switching schemes happen during that phase.

As an example, **V:C** determines a category of electrostatic harvesters in which the capacitance of C_V changes under voltage-constraint scheme during the recovery phase and under charge-constraint scheme during the harvesting phase.

2.1.5.2   Circuit Feature Based Categories

Each energy conversion cycle of a switching electrostatic harvester consists of four phases of operation as explained in Sect. 2.1.4. The moment that the capacitance of the variable capacitor is maximal ($C_V = C_{max}$) is the start of consecutive investment and harvesting phases, and the moment that $C_V = C_{min}$ is the start of consecutive reimbursement and recovery phases. Therefore, two main switching events occur at the start of the investment and reimbursement phases when $C_V = C_{max}$ and $C_V = C_{min}$, respectively. Switches are used to implement these two switching events, and a control circuit generates the gate pulses for these switches. An important circuit feature is that whether the gate pulses should be synchronized with these moments (when $C_V = C_{max}$ and $C_V = C_{min}$) or not. The naming convention based on this feature is suggested in this book with the following considerations:

- A switching event is synchronous if it should be synchronized with the moments that $C_V = C_{max}$ or $C_V = C_{min}$ and is asynchronous if it does not need to be synchronized with these moments.
- The letter "S" represents a synchronous event, and the letter "A" represents an asynchronous event.
- The naming includes two parts representing the switching event for the moment that $C_V = C_{max}$ and the switching event for the moment that $C_V = C_{min}$. These parts are separated with ":".
- The letters before ":" stand for the switching event at when $C_V = C_{max}$ and letters after ":" stand for the switching event at when $C_V = C_{min}$.

As an example, **A:S** determines a category of electrostatic harvesters in which the switching event for the start of the investment phase is asynchronous and the switching event for the start of the reimbursement phase is synchronous.

2.1.5.3   The Relation Between These Naming Conventions

There is a relation between categorizing based on the QV diagram and categorizing based on the circuit feature in the above. A diode may be employed as the switch to start the reimbursement phase when an electrostatic harvester follows a voltage-constraint (or no-constraint) in its harvesting phase. The diode can be employed in this scenario since it simply starts conducting when the voltage across C_V becomes equal to the voltage across the storage component. Similarly, a diode may be employed to start the investment phase when an electrostatic harvester follows a voltage-constraint (or no-constraint) scheme in its recovery phase. A diode is an asynchronous switch; therefore, the letters "V," "N," "CV," or "CN" in the QV based naming convention translate to "A" in the circuit feature based naming convention. However, the switches that start the reimbursement or the investment phases cannot be asynchronous if the harvester follows charge-constraint scheme in its harvesting or recovery phases. Therefore, the letter "C" in the QV based naming convention translates to "S" in the circuit feature based naming convention.

2.2   Generated Energy in a Variable Capacitor

During a full energy conversion cycle, an energy source changes the capacitance of a variable capacitor (C_V) from its maximal value (C_{max}) to its minimal value (C_{min}) and then back to its maximal value again. This cycle repeats, and an amount of energy converts into electrical form in each cycle. Depending on the circuit structure of the harvester, a part or all of the converted energy could be transferred to the storage component or the load. This energy is called the deliverable energy and is noted as E_{del} throughout this book. Due to the conduction losses, a part of this energy is lost before being delivered to the load or to the storage component.

The net energy that is delivered to the load or the storage component is called the net generated energy and is noted as E_{net} throughout this book. In an electrostatic harvester, E_{net} is always less than E_{del} due to the conduction losses. In this section, calculation of E_{del} is explained based on common QV diagrams for electrostatic harvesters. The calculation of E_{net} is discussed in more detail in Sect. 2.3.

In the following examples, the variable capacitor is initially charged to an arbitrary voltage, V_r. In practice, V_r is determined based on the imposed limitations by the maximum voltage across C_V, the maximum voltage that the employed switches can tolerate, and the amount of available energy in the energy source.

2.2.1   C:C Example

Figure 2.7 shows an electrostatic harvester in the category of C:C. A possible implementation of this QV diagram is the harvester circuit that is shown in Fig. 2.3. The operation of this harvester is explained in the following phases, from the QV diagram point of view:

Investment phase (a–b): At the beginning of this phase, the voltage across C_V is zero and the QV diagram is at point **a**. The variable capacitor gets charged to an initial voltage (V_r) during this phase, while the capacitance of C_V is maximal: $C_V = C_{max}$. Therefore, the diagram should move on the line with the slope equal to C_{max} and point **b** shows the end of this phase. The amount of energy that is received by C_V during this phase (E_{vi}) is

Fig. 2.7 The QV diagram for an electrostatic harvester in the C:C category

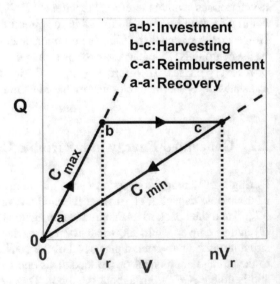

$$E_{vi} = \frac{1}{2}C_{max}V_r^2 - \frac{1}{2}C_{max}0^2 \Rightarrow E_{vi} = \frac{1}{2}C_{max}V_r^2. \tag{2.4}$$

Harvesting phase (b–c): The capacitance of C_V is maximal ($C_V = C_{max}$), and C_V is charged to V_r at the beginning of this phase; hence, point **b** shows the start of this phase. C_V gets isolated from the rest of the circuit, and an energy source changes the capacitance of C_V from C_{max} to C_{min}. Since C_V is isolated during this phase, the charge on C_V does not change; hence,

$$Q = CV \Rightarrow \begin{cases} n = \dfrac{C_{max}}{C_{min}} \\ Q \text{ is constant} \end{cases} \Rightarrow V_{Cv} : V_r \rightarrow nV_r. \tag{2.5}$$

During this phase, C_V harvests the following amount of energy (E_{hC}) from the energy source.

$$E_{hC} = \frac{1}{2}C_{min}(nV_r)^2 - \frac{1}{2}C_{max}V_r^2 \Rightarrow E_{hC} = \frac{1}{2}C_{max}(n-1)V_r^2. \tag{2.6}$$

E_{hC} in the above expression shows the amount of energy that an electrostatic generator harvests from the energy source under charge-constraint scheme in each cycle, when C_V is charged to V_r at the beginning of the harvesting phase. The ratio n is defined in (2.5).

Reimbursement phase (c–a): Point **c** shows the start of this phase, when the capacitance of C_V is minimal ($C_V = C_{min}$) and the voltage across C_V is nV_r. All the charge on C_V transfers to a load or an energy storage component (a battery or a large capacitor) during this phase. The voltage across C_V is zero at the end of this phase, and point **a** refers to this moment. The amount of energy that is obtained from C_V during this phase is

$$E_{vr} = \frac{1}{2}C_{min}(nV_r)^2 - \frac{1}{2}C_{min}0^2 \Rightarrow E_{vr} = \frac{1}{2}C_{min}(nV_r)^2. \tag{2.7}$$

Recovery phase (a–a): C_V is fully discharged during the reimbursement phase: the voltage across C_V is zero at the beginning of the recovery phase. Point **a** shows the start of this phase. C_V is isolated from the rest of the circuit during this phase; hence, the voltage across C_V remains at zero. The same point **a** shows the end of this phase.

The variable capacitor initially receives an amount of energy (E_{vi}) from the energy storage component and reimburses an amount of energy (E_{vr}) to this component (or the load) later. Therefore, the amount of deliverable energy from C_V to the storage component (or the load) is

$$E_{del} = E_{vr} - E_{vi}. \tag{2.8}$$

Using (2.4) and (2.7), E_{del} for an electrostatic harvester that follows the QV diagram in Fig. 2.7 is

$$E_{del} = \frac{1}{2}C_{min}(nV_r)^2 - \frac{1}{2}C_{max}V_r^2 \Rightarrow E_{Cv} = \frac{1}{2}C_{max}(n-1)V_r^2. \tag{2.9}$$

Generally, the amount of deliverable energy to the storage component, E_{del}, is not equal to the amount of the harvested energy in C_V during the harvesting phase, E_{hC}. However, E_{del} (as calculated in the above) and E_{hC} (as calculated in (2.6)) are equal for this harvester as all the energy in C_V is reimbursed to the storage component during the reimbursement phase.

In the above, E_{del} is found based on (2.8) and calculating the parameters E_{vi} and E_{vr}. However, E_{del} is directly calculable from the QV diagram. The enclosed area in the QV diagram of this harvester in Fig. 2.7 is equal to E_{del} in (2.9).

2.2.2   C:V Example 1

Figure 2.8 shows an electrostatic harvester in the category of C:V. The operation of this harvester during the investment and recovery phases is the same as explained for the C:C example in Fig. 2.7. The operation of this harvester during the harvesting and reimbursement phases is explained as follows.

Harvesting phase (b–c): At the beginning of this phase, the capacitance of C_V is maximal ($C_V = C_{max}$) and C_V is charged to V_r; hence, point **b** represents this moment. An energy source changes the capacitance of C_V from C_{max} to C_{min}.

Fig. 2.8 The QV diagram for an electrostatic harvester in the C:V category

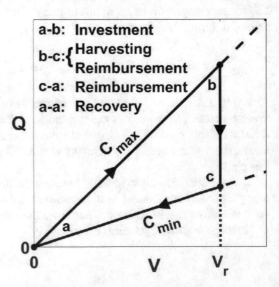

During this phase, C_V connects to the storage component with the voltage V_r across it. Therefore, the voltage across C_V is kept constant during this phase. Point **c** on the line with the slope of C_{min} marks the end of this phase and:

$$Q = CV \Rightarrow \begin{cases} Q_b = C_{max} V_r \\ V \text{ is constant} \end{cases} \Rightarrow Q_{Cv} : Q_b \to \frac{1}{n} Q_b. \tag{2.10}$$

During this phase, C_V harvests an amount of energy from the energy source. The calculation of this energy needs an insight of what happens in the circuit side, since C_V is connected to a battery (or a large capacitor) during this phase. This calculation is detailed in Sect. 4.1.4. The result is shown in below:

$$E_{hV} = \frac{1}{2} (C_{max} - C_{min}) V_r^2 \Rightarrow E_{hV} = \frac{1}{2} C_{min} (n - 1) V_r^2. \tag{2.11}$$

E_{hV} in the above expression shows the amount of energy that an electrostatic generator harvests from the energy source under voltage-constraint scheme in each cycle, when C_V is charged to V_r at the beginning of the harvesting phase. The ratio n is defined in (2.5).

Reimbursement phase (b–c and c–a): Line **b–c** in Fig. 2.8 represents both the harvesting and reimbursement phases, since C_V is connected to a load or the storage component during this time. According to (2.11), C_V harvests an amount of energy along this line. Simultaneously, the stored charge in C_V reduces n times based on (2.10) and this charge transfers to the storage component. Therefore, the amount of energy that is transferred from C_V during this phase is

$$E_{vr-bc} = (Q_b - Q_c) V_r \Rightarrow E_{vr-bc} = (C_{max} - C_{min}) V_r^2. \tag{2.12}$$

Not all the energy in C_V is transferred to the storage component along line **b-c** in the QV diagram, and at point **c**, an amount of energy is still remained in C_V. To transfer this remained energy to the storage component, C_V connects to the storage component through another set of switches, and the voltage across C_V changes from V_r to zero, while $C_V = C_{min}$; hence,

$$E_{vr-ca} = \frac{1}{2} C_{min} V_r^2 - \frac{1}{2} C_{min} 0^2 \Rightarrow E_{vr-ca} = \frac{1}{2} C_{min} V_r^2. \tag{2.13}$$

The amount of energy that is transferred to the storage component during the reimbursement phase is

$$E_{vr} = E_{vr-bc} + E_{vr-ca} \Rightarrow E_{vr} = (C_{max} - C_{min}) V_r^2 + \frac{1}{2} C_{min} V_r^2. \tag{2.14}$$

The energy, E_{vi}, for this harvester is the same as expressed in (2.4); hence, E_{del} for this harvester is calculated based on (2.8) as follows:

$$E_{del} = E_{vr} - E_{vi} = (C_{max} - C_{min})\,V_r^2 + \frac{1}{2}C_{min}V_r^2 - \frac{1}{2}C_{max}V_r^2$$

$$\Rightarrow E_{del} = \frac{1}{2}C_{min}(n-1)V_r^2. \tag{2.15}$$

In this harvester (same as the example for C:C), all the energy in C_V is reimbursed to the storage component during the reimbursement phase. Therefore, E_{del} in the above is equal to E_{hV} (as expressed in (2.11)) for this harvester. Same as before, calculating the enclosed area in the QV diagram of Fig. 2.8 shows the same amount of energy as E_{del} in the above.

2.2.3 C:V Example 2

Figure 2.9 shows another example of electrostatic harvesters in the C:V category. In the previous example, a set of switches other than those used during the harvesting phase are necessary to transfer the remained energy in C_V to the storage component at the end of the harvesting phase. In this example, the circuit topology does not include this set of switches, and hence, the remained energy in C_V is not transferred to the storage component. Therefore, E_{vr} in this example is

$$E_{vr} = E_{vr-bc} = (C_{max} - C_{min})\,V_r^2. \tag{2.16}$$

Point **c** shows the end of the harvesting and reimbursement phases in Fig. 2.9. The recovery phase starts at this point when $C_V = C_{min}$, and the voltage across C_V

Fig. 2.9 The QV diagram for an electrostatic harvester in the C:V category

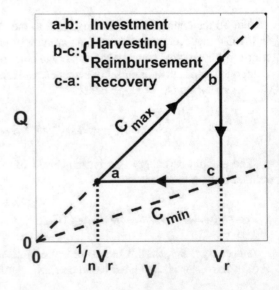

is V_r. The variable capacitor is isolated from the rest of the circuit, and the energy source changes the capacitance of C_V from C_{min} to C_{max}. Therefore, point **a** shows the end of the recovery phase where the voltage across C_V is n times less than V_r.

Point **a** marks the beginning of the investment phase. The amount of the energy that is transferred to C_V during the investment phase for this example is different to the previous examples, as the voltage across C_V is not zero at the beginning of this phase. Based on Fig. 2.9, E_{vi} for this harvester is

$$E_{vi} = \frac{1}{2}C_{max}V_r^2 - \frac{1}{2}C_{max}\left(\frac{1}{n}V_r\right)^2 \Rightarrow E_{vi} = \frac{1}{2}C_{max}\left(1 - \frac{1}{n^2}\right)V_r^2. \quad (2.17)$$

According to the above expressions for E_{vi} and E_{vr}, the amount of deliverable energy to the storage component in this harvester is

$$E_{del} = E_{vr} - E_{vi} = (C_{max} - C_{min})V_r^2 - \frac{1}{2}C_{max}\left(1 - \frac{1}{n^2}\right)V_r^2$$

$$\Rightarrow E_{del} = \frac{1}{2}C_{min}\left(n - 2 + \frac{1}{n}\right)V_r^2 = \frac{1}{2}C_{min}\frac{(n-1)^2}{n}V_r^2. \quad (2.18)$$

The amount of energy that C_V harvests during the harvesting phase in this harvester (E_{hV}) is the same as expressed in (2.11). In this harvester, E_{del} in the above is not equal to E_{hV}, since not all the energy in C_V is transferred to the storage component in this harvester. Calculating the enclosed area in Fig. 2.9 confirms that E_{del} in the above is equal to this area.

2.3 Net Generated Energy and Conduction Losses

The harvested and deliverable energies in a variable capacitor, E_h and E_{del}, are explained and calculated in Sect. 2.2. In this section, the net generated energy in an electrostatic harvester (E_{net}) is discussed. Due to the conduction losses in the investment and reimbursement circuitries, E_{net} is always less than E_{del}. The following two approaches may be followed to calculate E_{net}. The second approach is the preferred solution to calculate this energy in the rest of this book as it involves dealing with less parameters.

The following two approaches are explained for both non-sustainable and sustainable systems in Figs. 2.1 and 2.2. The storage component in the non-sustainable system of Fig. 2.1 is depicted as a battery, and this component is depicted as a reservoir capacitor (C_r) in the sustainable system of Fig. 2.2. However, it is possible to choose a battery or a large capacitor as the storage component for any of these systems. In the following approaches, the net generated energy or the conduction losses are calculated for any of these systems based on the voltage across the storage component. To generalize the analyses, the voltage across the storage component (whether it is a battery or a reservoir capacitor) is noted as V_{ri}. It is reasonable to assume that the variations in this voltage are negligible when a battery or a large capacitor is chosen. An analysis technique is presented in Sect. 2.4, for the cases that this variation should be taken into account.

2.3.1 First Solution

This approach involves finding the conduction losses in the investment and reimbursement circuitries directly.

2.3.1.1 Sustainable System

The reservoir capacitor (C_r) in Fig. 2.2 is initially charged to V_{ri} before the harvester starts operating. C_r charges the variable capacitor through the investment circuitry before the start of the harvesting phase. Later, the variable capacitor reimburses C_r through the reimbursement circuitry at the end of the harvesting phase. An amount of energy is lost in the investment and reimbursement circuitries in this process. This loss is due to the conduction losses in the components of these circuitries. To obtain the net generated energy in C_r, the following expression is used:

$$E_{net} = E_{del} - E_{li} - E_{lr}, \tag{2.19}$$

where E_{li} and E_{lr} are the conduction losses in the investment and reimbursement circuitries, respectively. These losses are calculated using the following expressions:

$$E_{li}(t) = R_i i_{iR}^2(t)t + V_{iS} i_{iS}(t)t \Rightarrow E_{li} = R_i \int i_{iR}^2 dt + V_{iS} \int i_{iS} dt$$

$$E_{lr}(t) = R_r i_{rR}^2(t)t + V_{rS} i_{rS}(t)t \Rightarrow E_{lr} = R_r \int i_{rR}^2 dt + V_{rS} \int i_{rS} dt, \tag{2.20}$$

where R_i and R_r are the equivalent resistances of the conducting paths in the investment and reimbursement circuitries; i_{iR} and i_{rR} are the currents that go through these resistances. V_{iS} and V_{rS} are the forward voltage drops across the diodes in the investment and reimbursement circuitries; i_{iS} and i_{rS} are the currents that go through these switches.

2.3.1.2   Non-Sustainable System

The only difference between the non-sustainable system in Fig. 2.1 and the sustainable system in Fig. 2.2 is that the reimbursement circuitry is placed between the variable capacitor and the load. Subsequently, the generated energy in C_V is transferred to the load, instead of the storage component. Therefore, all the above expressions are applicable to the non-sustainable system.

2.3.1.3   Overall

In the above expressions, R_i, R_r, V_{iS}, and V_{rS} are the parameters of the conducting paths. Other than these parameters, the integral of the currents and the integral of the squared currents that go through each of these components impact the conduction losses. Therefore in this approach, to calculate E_{net} in (2.19), the integral of the currents and the integral of the squared currents should be found.

2.3.2   Second Solution

In this solution, the conduction losses in the reimbursement and investment circuitries do not need to be calculated directly to find the net generated energy.

2.3.2.1   Sustainable System

The reservoir capacitor, C_r, connects to the variable capacitor, C_V, through the investment circuitry to initially charge C_V. This charge transfer occurs when $C_V = C_{max}$ and during the investment phase. Due to conduction losses, not all of the energy obtained from C_r is transferred to C_V. Hence, the amount of energy that is received by C_V during this phase (E_{vi}) is expressed as follows:

$$E_{vi} = E_{ri} - E_{li}, \tag{2.21}$$

where E_{ri} and E_{li} are the amount of energy that is obtained from C_r and the conduction losses during this phase, respectively.

The energy harvesting phase starts after the investment phase. During this phase, the energy source changes the capacitance of the variable capacitor from the maximal value, C_{max}, towards the minimal value, C_{min}. As a result of this change, an amount of energy is transduced from another form into electrical form and stored in the variable capacitor.

At the end of the harvesting phase or during this phase, C_V connects to C_r through reimbursement circuitry to store the generated energy in C_r. Not all the energy that is obtained from the variable capacitor (E_{vr}) is transferred to C_r. Hence,

$$E_{vr} = E_{rr} + E_{lr}, \tag{2.22}$$

where E_{rr} is the amount of energy that is received by C_r and E_{lr} is the amount of conduction losses during this phase.

The recovery phase starts after the reimbursement phase. During this phase, the energy source changes capacitance of the variable capacitor from minimal value, C_{min} towards maximal value, C_{max}. At the end of this phase, the variable capacitor becomes ready for the next operating cycle.

The net generated energy is the amount of energy that is stored in C_r during a full cycle, consisting of all the above phases:

$$E_{net} = E_{rr} - E_{ri}, \tag{2.23}$$

where E_{rr} is the amount of energy that is reimbursed to C_r during the reimbursement phase, and E_{ri} is the amount of energy that C_r invests to C_V during the investment phase. In Fig. 2.2, i_{ri} and i_{rr} are the investment current and the reimbursement current, respectively, that go through C_r. E_{rr} and E_{ri} are calculated in the following steps. Considering that V_{ri} is the voltage across C_r at the beginning of the investment phase and V_{rii} is the voltage across this capacitor at the end of this phase, E_{ri} is calculated as follows:

$$V_{rii} = V_{ri} - \frac{1}{C_r} \int i_{ri} dt \tag{2.24}$$

$$\Rightarrow E_{ri} = \frac{1}{2} C_r \left(V_{ri}^2 - V_{rii}^2 \right)$$

$$\Rightarrow E_{ri} = \frac{1}{2} C_r \left(V_{ri}^2 - \left(V_{ri} - \frac{1}{C_r} \int i_{ri} dt \right)^2 \right)$$

$$\Rightarrow E_{ri} = \frac{1}{2} \int i_{ri} dt \left(2V_{ri} - \frac{1}{C_r} \int i_{ri} dt \right). \tag{2.25}$$

In case of a negligible leakage for C_r, the voltage across this capacitor at the beginning of the reimbursement phase will be equal to V_{rii}. At the end of the reimbursement phase, the voltage across this capacitor (V_{rir}) is as follows:

$$V_{rir} = V_{rii} + \frac{1}{C_r} \int i_{rr} dt \Rightarrow V_{rir} = V_{ri} - \frac{1}{C_r} \int i_{ri} dt + \frac{1}{Cr} \int i_{rr} dt. \tag{2.26}$$

Knowing V_{rir} and V_{rii}, E_{rr} is obtained:

$$E_{rr} = \frac{1}{2} C_r \left(V_{rir}^2 - V_{rii}^2 \right)$$

$$\Rightarrow E_{rr} = \frac{1}{2}C_r\left(\left(V_{ri} - \frac{1}{C_r}\int i_{ri}dt + \frac{1}{Cr}\int i_{rr}dt\right)^2 - \left(V_{ri} - \frac{1}{C_r}\int i_{ri}dt\right)^2\right).$$

(2.27)

Using the expressions in (2.23), (2.25), and (2.27), the net generated energy is expressed as follows:

$$E_{net} = \frac{1}{2}\left(\int i_{rr}dt - \int i_{ri}dt\right)\left(2V_{ri} + \frac{1}{C_r}\left(\int i_{rr}dt - \int i_{ri}dt\right)\right).$$ (2.28)

In the above expression, the net generated energy is expressed based on C_r, V_{ri}, the integral of i_{ri}, and the integral of i_{rr}. In Sect. 2.2, it is explained that C_r and the initial voltage across it (V_{ri}) are determined by the application. Therefore, E_{net} is obtained for any circuit topology by calculating the integral of i_{ri} and the integral of i_{rr} for different switching schemes.

Based on (2.21), (2.22), and (2.23), the total conduction losses during the investment and reimbursement phases ($E_{loss-tot}$) are expressed as follows:

$$E_{loss-tot} = E_{li} + E_{lr} = (E_{ri} - E_{vi}) + (E_{vr} - E_{rr})$$

$$\Rightarrow E_{loss-tot} = (E_{vr} - E_{vi}) - (E_{rr} - E_{ri})$$

$$\Rightarrow E_{loss-tot} = E_{del} - E_{net}.$$ (2.29)

As discussed in Sect. 2.2, E_{del} is the deliverable energy to the storage component and may be calculated either directly from the QV diagram or knowing the values of E_{vr} and E_{vi} and using (2.8). The net delivered energy to the storage component, E_{net}, is always less than E_{del} due to the conduction losses in the investment and reimbursement paths. Therefore, the total conduction losses are calculated by knowing E_{net} and E_{del} based on the above expression.

2.3.2.2   Non-Sustainable System

To obtain the net generated energy in this type of harvester, the expression in (2.23) is used. Therefore, E_{rr} and E_{ri} should be calculated for the structure in Fig. 2.1. The current that goes through the storage component during the investment phase (i_{ri}) and the current that goes through the load during the reimbursement phase (i_{rr}) are specified in this figure. In this system, E_{ri} is equal to

$$E_{ri} = V_B \int i_{ri}dt,$$ (2.30)

where V_B is the voltage of the battery in Fig. 2.1. The current that goes to a capacitor from a battery and its integral is discussed in more detail in Sects. 3.1 and 4.1, depending on the circuitry schematic. In case that a reservoir capacitor (C_r) is used for the storage component, the expression for E_{ri} would be the same as derived for the sustainable system in (2.25).

In this system, E_{rr} is not the amount of energy that is transferred back to the storage component and refers to the amount of energy that is transferred to the load (or C_s) during the reimbursement phase. In the case that a storage capacitor is used, E_{rr} is obtained as follows:

$$V_{sf} = V_s + \frac{1}{C_s} \int i_{rr} dt$$

$$\Rightarrow E_{rr} = \frac{1}{2} C_s \left(V_{sf}^2 - V_s^2 \right)$$

$$\Rightarrow E_{rr} = \frac{1}{2} \int i_{rr} dt \left(2V_s + \frac{1}{C_s} \int i_{rr} dt \right). \tag{2.31}$$

In the case that a storage capacitor does not exist and the generated energy in C_V transfers to the load directly, E_{rr} would depend on the type of the load. If the load is capacitive, E_{rr} is calculated similar to the above, where V_s and C_s are replaced with V_L and C_L (C_L is the load capacitor with a voltage of V_L across it). For a resistive load, E_{rr} is calculated as follows:

$$E_{rr} = R_L \int i_{rr}^2(t) dt. \tag{2.32}$$

Knowing E_{ri} and E_{rr}, E_{net} is calculated based on (2.23). The conduction losses are then calculated from (2.29), where E_{del} is obtained from the QV diagram of the harvester.

2.3.2.3 Overall

In this solution, the net generated energy (E_{net}) is obtained based on the integral of the currents that go through the investment and reimbursement circuitries. Therefore, calculating the conduction losses in the investment and reimbursement circuitries separately is not necessary in finding E_{net}. Instead, the total conduction losses in the investment and reimbursement paths ($E_{loss-tot}$) are calculated based on E_{net} and E_{del} (which is calculable from the QV diagram). The calculation of E_{net} and $E_{loss-tot}$ does not depend on the integral of the squared currents (in the investment and reimbursement paths) in this solution. The only exception to this is for the non-sustainable system with resistive load. As can be seen in (2.32), E_{rr} depends on the integral of the squared i_{rr} for this case. However, calculating E_{rr} for this case is straightforward. This is shown in finding E_{rr} in Sect. 2.3.3.2. Therefore, the calculation of E_{net} and $E_{loss-tot}$ is simpler in this solution compared to the first solution. However, the conduction losses are calculated totally ($E_{loss-tot}$), and the values for the conduction losses in the investment and reimbursement circuitries (E_{li} and E_{lr}) are not obtained separately in this solution.

2.3.3   Example

In this section, the first and second solutions to find the net generated energy and the conduction losses are applied on the circuit-level example in Fig. 2.3. The QV diagram of this circuit is shown in Fig. 2.7 and is explained in Sect. 2.2.1.

2.3.3.1   First Solution

The QV diagram of this harvester is the same as in Fig. 2.7, where V_r is replaced with V_B. The deliverable energy to the load is obtained by calculating the enclosed area in this figure as follows:

$$E_{del} = \frac{1}{2} C_{max} (n - 1) V_B{}^2. \tag{2.33}$$

The current that goes through the battery, S_1, and C_V during the investment phase is equal to

$$i_{ri}(t) = \frac{V_B}{R_{S1}} e^{-\frac{t}{\tau_i}}, \qquad \tau_i = C_{max} R_{S1}, \tag{2.34}$$

where R_{S1} is the on-resistance of S_1. The conduction losses during this phase are calculated as follows:

$$E_{li} = R_{S1} \int i_{ri}(t)^2 dt$$

$$\Rightarrow E_{li} = R_{S1} \frac{V_B{}^2}{R_{S1}{}^2} \int_0^\infty e^{-\frac{2t}{\tau_i}} dt$$

$$\Rightarrow E_{li} = \frac{1}{2} C_{max} V_B{}^2, \tag{2.35}$$

where it is assumed that S_1 is on for long enough that $i_{ri}(t)$ reaches to zero. As can be seen, the conduction losses in this case do not depend on R_{S1}. This is discussed in more detail in Chap. 3.

The current that goes through C_V, S_2, and R_L during the reimbursement phase is equal to

$$i_{rr}(t) = \frac{n V_B}{R_{S2} + R_L} e^{-\frac{t}{\tau_r}}, \qquad \tau_r = C_{min} (R_{S2} + R_L), \tag{2.36}$$

where R_{S2} is the on-resistance of S_2. The conduction losses during the reimbursement phase are calculated as follows:

$$E_{lr} = R_{S2} \int i_{rr}(t)dt$$

$$\Rightarrow E_{lr} = R_{S2} \frac{n^2 V_B^2}{(R_{S2} + R_L)^2} \int_0^\infty e^{-\frac{2t}{\tau_r}} dt$$

$$\Rightarrow E_{lr} = \frac{1}{2} C_{min} n^2 V_B^2 \frac{R_{S2}}{R_{S2} + R_L}, \tag{2.37}$$

where it is assumed that S_2 is on for long enough that $i_{rr}(t)$ reaches to zero.

Using the above values for E_{del}, E_{li}, and E_{lr}, the expression in (2.19) is used to find the net generated energy:

$$E_{net} = E_{del} - E_{li} - E_{lr}$$

$$\Rightarrow E_{net} = \frac{1}{2} C_{max} V_B^2 \left(\frac{R_L}{R_L + R_{S2}} n - 2 \right). \tag{2.38}$$

The total conduction losses are obtained as follows:

$$E_{loss-tot} = E_{li} + E_{lr} = \frac{1}{2} C_{max} V_B^2 \left(1 + \frac{R_{S2}}{R_{S2} + R_L} n \right). \tag{2.39}$$

In the above expressions for $i_{ri}(t)$ and $i_{rr}(t)$, it is assumed that $\tau_i, \tau_r \ll T_V$, where T_V is defined in (2.1). Therefore, the changes in the capacitance of the variable capacitor are negligible during the period that i_{ri} or i_{rr} reaches to zero. This condition is evident in Fig. 2.4.

2.3.3.2   Second Solution

The amount of energy that is obtained from battery during the investment phase is equal to

$$E_{ri} = V_B \int i_{ri}(t)dt$$

$$\Rightarrow E_{ri} = V_B \frac{V_B}{R_{S1}} \int_0^\infty e^{-\frac{t}{\tau_i}} dt$$

$$\Rightarrow E_{ri} = C_{max} V_B^2, \tag{2.40}$$

where $i_{ri}(t)$ is expressed in (2.34). The energy that is received by the load during the reimbursement phase is calculated as follows:

$$E_{rr} = R_L \int i_{rr}^2(t)dt$$

$$\Rightarrow E_{rr} = R_L \frac{n^2 V_B^2}{(R_L + R_{S2})^2} \int_0^\infty e^{-\frac{2t}{t_r}} dt$$

$$\Rightarrow E_{rr} = \frac{1}{2} C_{min} n^2 V_B^2 \frac{R_L}{R_L + R_{S2}}, \tag{2.41}$$

where i_{rr} is expressed in (2.36). Using (2.23) and the above values for E_{ri} and E_{rr}, the net generated energy is calculated:

$$E_{net} = E_{rr} - E_{ri} = \frac{1}{2} C_{max} V_B^2 \left(\frac{R_L}{R_L + R_{S2}} n - 2 \right). \tag{2.42}$$

As expected, the result is the same as in (2.38). The total conduction losses in this solution are calculated by using (2.29). The values of E_{del} and E_{net} in this expression are found in (2.33) and (2.42), respectively. The result is the same as in (2.39).

For this harvester $R_L \gg R_{S2}$ in practice, therefore,

$$E_{net} = \frac{1}{2} C_{max} V_B^2 (n - 2). \tag{2.43}$$

In case that $R_L \gg n R_{S2}$, the total conduction losses are as below for this harvester:

$$E_{loss-tot} = \frac{1}{2} C_{max} V_B^2. \tag{2.44}$$

2.4 Discrete Analysis of Electrostatic Harvesters

In practice, the voltages across the components of an electrostatic harvester are different in each operating cycle due to a likely difference between the amount of generated and consumed energy. However in the above sections, the operation of electrostatic harvesters and the energy calculations are explained and derived based on the assumption that the variable capacitor charges to the same voltage at the beginning of the harvesting phase in each cycle. This assumption is valid in many applications that either the storage components are chosen to be much bigger than the capacitances of the variable capacitor or a power management block is used to balance the generated and consumed energies.

In several occasions, it is quite necessary to consider the evolution of the voltages across the components of an electrostatic harvester. This includes the cases where the storage components are not much bigger than the capacitances of the variable capacitor. This may be the case when the storage components are chosen to be integrated capacitors. The other example is finding the voltages during the start-up period, during which the voltages have not yet got to a constant level.

In this section, a discrete analysis for the cases that the evolution of the voltages should be considered is introduced. The method is then applied for a few common occasions. More sophisticated examples where this analysis is used are discussed in Chap. 3.

2.4.1 Discrete Analysis

In an electrostatic harvester, a full operating cycle consists of four phases as explained in Sect. 2.1.1. The same sequence of events occur in each cycle, and this cycle repeats with a frequency of f_V, as expressed in (2.1). Following the sequence of events in a cycle, the voltage across any components after a cycle may be expressed as a function of the voltages across all the components before that cycle. This way, the same relation is valid for the next cycle, since the same sequence of events occur in each cycle. As an example, suppose the following equation is obtained for a harvester by going through the sequence of events in a cycle:

$$V_{Cr}(3) = \alpha_1 V_{Cr}(2) + \beta_1 V_{Cv}(2) + a_1, \tag{2.45}$$

where V_{Cr} and V_{Cv} are the voltages across the reservoir capacitor and the variable capacitor, respectively. In the above expression, the voltage across C_r in the third cycle is expressed based on the voltages across C_r and C_V in the second cycle. In this harvester, the following equation is also correct:

$$V_{Cr}(4) = \alpha_1 V_{Cr}(3) + \beta_1 V_{Cv}(3) + a_1. \tag{2.46}$$

In the above equations, α_1, β_1, and a_1 are constant coefficients. These coefficients are defined by the capacitances and the voltages that are considered constant, e.g. the voltage across a battery or a large capacitor. The numbers in the parentheses refer to the cycle number. According to this example, if a relation is obtained for the third cycle in a harvester, the same expression is correct for the fourth cycle in that harvester. This recurrence is true for any cycle; hence, in the harvester of this example, the following recursive formula can be written:

$$V_{Cr}(k+1) = \alpha_1 V_{Cr}(k) + \beta_1 V_{Cv}(k) + a_1. \tag{2.47}$$

It is not possible to solve the above recursive formula yet, since it contains two unknown terms, i.e. $V_{Cr}(k+1)$ and $V_{Cv}(k)$. Following the events in a cycle, another relation between these two unknowns should be found. Based on these two expressions, two independent recursive expressions should be obtained for $V_{Cr}(k)$ and $V_{Cv}(k)$. A recursive expression for $V_{Cr}(k)$ is independent if it only contains $V_{cr}(k+z)$ terms, where $z \in \mathbb{Z}$ and it does not contain any $V_{Cv}(k+z)$ terms. Same rules apply for obtaining an independent recursive formula for $V_{Cv}(k)$. Therefore, the general independent recursive expressions for the above harvester are as follows:

$$\alpha_z V_{Cr}(k+z) + \alpha_{z-1} V_{Cr}(k+z-1) + \ldots + \alpha_1 V_{Cr}(k+1) + \alpha_0 V_{Cr}(k) = a_1,$$
(2.48)

$$\beta_z V_{Cv}(k+z) + \beta_{z-1} V_{Cv}(k+z-1) + \ldots + \beta_1 V_{Cv}(k+1) + \beta_0 V_{Cv}(k) = b_1.$$
(2.49)

Solving these independent recursive expressions, closed-form formulas can be obtained for $V_{Cr}(k)$ and $V_{Cv}(k)$. Using these formulas, discrete values for the voltages across the variable capacitor and the reservoir capacitor are obtained. These discrete values show the voltages across C_V and C_r at a specific moment, e.g. after the reimbursement phase, in each cycle. This analysis enables examining the evolution of different voltages in an electrostatic harvester that is useful for a range of different purposes. In the rest of this section, a method for solving recursive expressions is introduced. The reader then becomes familiar with the examples of the discrete analysis. The real-world applications of this analysis are detailed in Chap. 3.

2.4.2   Solving Linear Non-Homogeneous Recursive Expressions

Following the phases of operation in a cycle, an independent recursive formula, as in (2.48)–(2.49), can be written for the voltage of any node in an electrostatic harvester. As can be seen in these formulas, the general recursive expressions for electrostatic harvesters are linear and non-homogeneous. In this section, a method is introduced for solving these recursive formulas for electrostatic harvesters. This method is explained in detail for different possible cases in [1]. Using this method here, closed-form formulas are found for the voltage of any node in an electrostatic harvester, e.g. $V_{Cr}(k)$. This formula should satisfy the recursive form in (2.48).

The equation in (2.48) may be solved as follows:

$$V_{Cr}(k) = V_{Cr-g}(k) + V_{Cr-p}(k),$$
(2.50)

where $V_{Cr-g}(k)$ is a general response to (2.48) and $V_{Cr-p}(k)$ is a particular response to (2.48). To obtain the general response to (2.48), a_1 is considered to be zero. A possible response to this equation when $a_1 = 0$ is assumed as follows:

$$V_{Cr}(k) = r^k, \quad r \in \mathbb{R}.$$
(2.51)

Replacing the above response in (2.48) where $a_1 = 0$, the below steps are followed to obtain the general response:

$$\alpha_z r^{k+z} + \alpha_{z-1} r^{k+z-1} + \ldots + \alpha_1 r^{k+1} + \alpha_0 r^k = 0, \tag{2.52}$$

$$\Rightarrow r^k \left(\alpha_z r^z + \alpha_{z-1} r^{z-1} + \ldots + \alpha_1 r^1 + \alpha_0 \right) = 0. \tag{2.53}$$

The equation in the parentheses of (2.53) is called the characteristic equation and has z roots. Assuming all the roots are distinct (not repeated), the general response to (2.48) is obtained as follows:

$$V_{Cr-g}(k) = d_z r_z^{\ k} + d_{z-1} r_{z-1}^{\ k} + \ldots + d_1 r_1^{\ k}, \tag{2.54}$$

where $d_z, d_{z-1}, \ldots, d_1, d_0$ are the constant coefficients. To find a particular response to (2.48), the following equation is considered:

$$V_{Cr-p}(k) = \frac{a_1}{\alpha_z r^z + \alpha_{z-1} r^{z-1} + \ldots + \alpha_1 r^1 + \alpha_0}, \tag{2.55}$$

where r should be replaced with one, since a_1 is a constant and the particular response would be a constant too. Therefore,

$$V_{Cr-p}(k) = \frac{a_1}{\alpha_z + \alpha_{z-1} + \ldots + \alpha_1 + \alpha_0}. \tag{2.56}$$

According to (2.50) and the derived expressions in (2.54) and (2.56) for $V_{Cr-g}(k)$ and $V_{Cr-p}(k)$, the solution to (2.48) is

$$V_{Cr}(k) = d_z r_z^{\ k} + d_{z-1} r_{z-1}^{\ k} + \ldots + d_1 r_1^{\ k} + \frac{a_1}{\alpha_z + \alpha_{z-1} + \ldots + \alpha_1 + \alpha_0}, \tag{2.57}$$

where the coefficients are obtained by considering the initial conditions. Initial conditions in electrostatic harvesters refer to the initial voltages across the capacitors at the start of the operation of the harvester. In this book, $k = 0$ marks the first operating cycle. Therefore in obtaining the coefficients in the above expression, one formula should be the result for $k = 0$, where $V_{Cr}(0)$ denotes the initial voltage across C_r before the harvester starts working.

2.4.3   Charge-Depletion Study

In the non-sustainable harvesters (as depicted in Fig. 2.1), no energy is transferred back to the storage component. Therefore, the voltage across the storage component eventually decreases and reaches to zero. At this point, the initial charge in the storage component depletes and the harvester stops working. In this section, a case study is presented to illustrate the charge-depletion issue, and how using the discrete analysis is helpful in this case. Later, the net generated energy of this harvester is

Fig. 2.10 Typical
non-sustainable harvester
with reservoir capacitor

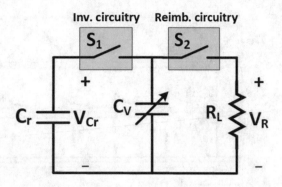

calculated using the discrete analysis. This calculation reveals the impact of the
harvester and a condition on the variable capacitor that should be satisfied for the
harvester to have a positive net generated energy.

2.4.3.1   Case Study

Figure 2.10 shows the typical non-sustainable harvester with a reservoir capacitor.
This circuit is the same as the circuit in Fig. 2.3 except that a reservoir capacitor is
used instead of a battery. Therefore, the operation of these two circuits is identical.
However, the voltage decrease across the battery in the circuit of Fig. 2.3 is not
identifiable in few operating cycles.

The voltages across the capacitors and the load of the circuit in Fig. 2.10 are
shown in Fig. 2.11. The decrease in all the voltages of this circuit is notifiable,
since the reservoir capacitor is selected relatively small in the circuit of Fig. 2.10
to illustrate the charge depletion in few operating cycles. The explanations in
Sect. 2.1.3 should be referred to recognize the phases of operation in a cycle for
this harvester. A closed-form formula for $V_{Cr}(k)$ is obtained as follows based on
these explanations.

The voltage across C_r at the kth cycle ($V_{Cr}(k)$) before the start of the investment
phase is marked in Fig. 2.11. The voltage across the variable capacitor is zero at
this moment. S_1 turns on, and C_r connects to C_V when $C_V = C_{max}$ to charge the
variable capacitor initially; therefore,

$$V_f = \frac{C_r V_{Cr}(k) + C_{max}.0}{C_r + C_{max}} \Rightarrow V_{Cr}(k) = \left(1 + \frac{C_{max}}{C_r}\right) V_f, \qquad (2.58)$$

where V_f is the final voltage across C_r and C_V at the end of this phase, based
on charge conservation law. This has been discussed in more detail in Sect. 3.1.5. S_1
turns off at the end of this phase and remains off for the rest of operating phases until
the start of next cycle. Therefore, the voltage across C_r at the start of the investment
phase at the $(k + 1)$th cycle ($V_{Cr}(k + 1)$) is the same as V_f. Replacing this value in
(2.58):

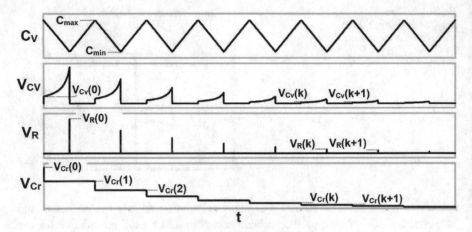

Fig. 2.11 The changes in the capacitance of the variable capacitor and the voltages across C_V, C_1, and R_L in the circuit of Fig. 2.10

$$V_{Cr}(k+1) = V_f \Rightarrow \left(1 + \frac{C_{max}}{C_r}\right) V_{Cr}(k+1) - V_{Cr}(k) = 0. \tag{2.59}$$

The above equation is an independent recursive formula for the voltage across C_r. This equation is a linear homogeneous recursive formula, and the method in Sect. 2.4.1 can be used for solving it. The characteristic equation for this formula is as follows:

$$\left(1 + \frac{C_{max}}{C_r}\right) r - 1 = 0 \Rightarrow r = \frac{C_r}{C_r + C_{max}}. \tag{2.60}$$

Knowing the only root of the characteristic equation from the above, the solution to (2.59) is as follows:

$$V_{Cr}(k) = d_1 \left(\frac{C_r}{C_r + C_{max}}\right)^k, \tag{2.61}$$

where d_1 is a constant coefficient and should be found based on the initial conditions. Replacing $k = 0$ in the above equation:

$$V_{Cr}(0) = d_1 \Rightarrow d_1 = V_r, \tag{2.62}$$

where the initial voltage across C_r ($V_{Cr}(0)$) is assumed to be V_r; therefore,

$$V_{Cr}(k) = V_r \left(\frac{C_r}{C_r + C_{max}}\right)^k. \tag{2.63}$$

The voltage across the variable capacitor is zero at the start of the investment phase at each cycle. Therefore, $V_{Cv}(k)$ is marked in Fig. 2.11 as the voltage across C_V at the end of this phase at the kth cycle. The voltage across C_V at this moment is equal to the voltage across C_r at the end of the investment phase; therefore,

$$V_{Cv}(k) = V_f \Rightarrow V_{Cv}(k) = V_{Cr}(k+1)$$

$$\Rightarrow V_{Cv}(k) = V_r \left(\frac{C_r}{C_r + C_{max}} \right)^{k+1}, \tag{2.64}$$

where V_f is the final voltage across C_r and C_V at the end of the investment phase, and based on (2.59), this voltage is the same as $V_{Cr}(k+1)$. From the above expression, it can be confirmed that the voltage across C_V at the end of the investment phase of the first cycle is

$$V_{Cv}(0) = V_r \left(\frac{C_r}{C_r + C_{max}} \right). \tag{2.65}$$

In the steps of finding the above expression, $V_{Cv}(k)$ is considered to be the voltage across C_V at the end of the investment phase at kth cycle. Therefore, the above expression shows this voltage only at the end of the investment phase in each cycle, and it does not show any other variations in this voltage during a full cycle.

In finding a closed-form formula for any voltage in an electrostatic harvester, a recursive expression should be found by following the sequence of events in kth cycle. It is arbitrary to mark the voltage of interest at any moment of this cycle and follow the steps to find the voltage at the same moment in the next cycle. The resulting equation from solving this recursive formula then shows the voltage of interest at the selected moment in a cycle. However, knowing this voltage reveals other values of this voltage during a full operating cycle by following the sequence of events happening in a cycle.

The voltage across the load is marked at the start of the reimbursement phase at kth cycle in Fig. 2.11. At the start of the reimbursement phase, C_V connects to R_L. Therefore, the voltage across the load is equal to the voltage across C_V at this moment. This voltage is n times of $V_{Cv}(k)$ as a result of the harvesting phase, where $n = C_{max}/C_{min}$; therefore,

$$V_R(k) = n V_{Cv}(k) \Rightarrow V_R(k) = n V_r \left(\frac{C_r}{C_r + C_{max}} \right)^{k+1}. \tag{2.66}$$

In all the derived expressions for $V_{Cr}(k)$, $V_{Cv}(k)$, and $V_R(k)$, the term that is powered to k is equal to the root that is found in (2.60). This term is always less than 1, and therefore, $V_{Cr}(k)$, $V_{Cv}(k)$, and $V_R(k)$ eventually become zero. This term is closer to 1 for larger capacitance of C_r. In this case, the voltages become zero in slower pace.

Fig. 2.12 Simple connection
of a reservoir capacitor to a
resistive load

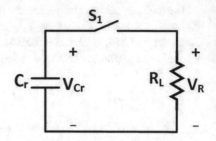

The charge depletion also exists in sustainable harvesters when the energy that is transferred back to the storage component (during the reimbursement phase) is equal to or less than the energy that is obtained from it (during the investment phase). The same method as introduced here may be used to identify this issue in sustainable harvesters. Two sustainable structures suffering from this issue are studied in Sect. 3.2.

2.4.3.2 Delivered Energy to the Load

To realize the impact of an electrostatic harvester being presented in a system, Fig. 2.12 shows simple connection of a reservoir capacitor to a resistive load. This circuit is in fact the circuit in Fig. 2.10 with no electrostatic harvester.

The voltages across the reservoir capacitor and the load for the circuit of Fig. 2.12 are shown in Fig. 2.13. At $t = t_s$, S_1 turns on and C_r connects to the load. The voltage across the load and the reservoir capacitor after this point is the same as in (2.2), where $\tau = C_r R_L$. After a duration about 5τ, the voltage across C_r reaches to almost zero. At this point, all the energy that was stored in C_r is already transferred to the load in the form of the current that has gone through R_L. Therefore, the energy that is received by the load is as below, provided that $R_{S1} \ll R_L$ (R_{S1} is the on-resistance of S_1). This is discussed in more detail in finding E_{rr} in Sect. 2.3.3.2:

$$E_{RL} = \frac{1}{2} C_r V_r^2. \tag{2.67}$$

The energy that is delivered to load in the circuit of Fig. 2.10 is different in each cycle. This is evident from Fig. 2.11, where the voltage across the variable capacitor at the start of the reimbursement phase decreases cycle by cycle. In this circuit, S_2 turns on at the start of the reimbursement phase and C_V connects to the load. The energy in the variable capacitor transfers to the load the same way as explained in the above for the simple connection of C_r to the load. However, the capacitance of the variable capacitor is much less than C_r, and the voltage across it is more than the voltage across C_r compared to the above case. The energy that is transferred to the load in the kth cycle is obtained as below, provided that $R_{S2} \ll R_L$ (R_{S2} is the on-resistance of S_2):

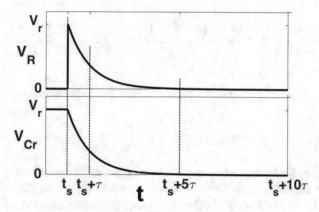

Fig. 2.13 The voltages across the reservoir capacitor and the load in the circuit of Fig. 2.12

$$E_{rr}(k) = \frac{1}{2}C_{min}(V_R(k))^2 \Rightarrow E_{rr}(k) = \frac{1}{2}C_{min}n^2V_r^2\left(\frac{C_r}{C_r + C_{max}}\right)^{2k+2},$$
(2.68)

where $V_R(k)$ is the voltage across the load at the start of the reimbursement phase, and this voltage is calculated in (2.66). The above equation is correct with the following assumption:

$$\tau_{Reimb.} \gg \frac{T_V}{2} \Rightarrow C_{min}R_L \gg \frac{T_V}{2},$$
(2.69)

where $\tau_{Reimb.}$ refers to the time constant of the reimbursement phase that depends on the values of C_{min} and R_L. This time constant is normally much less than the time period of the variable capacitor as defined in (2.1). Figure 2.11 confirms the validity of this assumption for the component values considered here. As can be seen in this figure, the time period of the variable capacitor is much larger than the time constant of the reimbursement phase.

The total energy that would be transferred to the load after the voltage across C_r fully depletes is obtained as follows:

$$E_{rr} = \sum_{k=0}^{k=\infty} E_{rr}(k) \Rightarrow E_{rr} = \frac{1}{2}C_{min}n^2V_r^2 \sum_{k=0}^{k=\infty}\left(\frac{C_r}{C_r + C_{max}}\right)^{2k+2},$$
(2.70)

where the sum of the geometric series in the above is calculated as follows:

$$\sum_{k=0}^{k=\infty} \left(\frac{C_r}{C_r + C_{max}} \right)^{2k+2} = \frac{\left(\frac{C_r}{C_r + C_{max}} \right)^2}{1 - \left(\frac{C_r}{C_r + C_{max}} \right)^2} = \frac{C_r{}^2}{C_{max} \left(2C_r + C_{max} \right)}. \qquad (2.71)$$

Therefore, replacing the above value in (2.70), E_{rr} is obtained as follows:

$$E_{rr} = \frac{1}{2} C_r V_r{}^2 \left(\frac{nC_r}{2C_r + C_{max}} \right). \qquad (2.72)$$

Comparing E_{rr} in the above expression with E_{RL} for connecting C_r directly to the load in (2.67), the following condition should be satisfied on n so that the harvester delivers more energy to the load compared to the circuit in Fig. 2.12:

$$nC_r > (2C_r + C_{max}) \Rightarrow n > \frac{2C_r + C_{max}}{C_r} \Rightarrow n > 2 + \frac{C_{max}}{C_r}. \qquad (2.73)$$

The harvester makes the situation worse than connecting C_r directly to the load, if the above condition is not satisfied. This is due to the losses during the investment phases. The conduction losses during the investment phase occur in each cycle, when the harvester is used. However, in connecting C_r directly to the load as depicted in Fig. 2.12, this occurs only once. Therefore, if the above condition is not satisfied for the value of n, the harvester should not be used since the net generated energy in the harvester would be negative in this case. The net generated energy and the conduction losses of this harvester are calculated in the following section.

The above analysis for the conditions on the value of n is applicable to the harvester circuit in Fig. 2.3, where the storage component is battery. To this end, the condition of $C_r \gg C_{max}$ may be assumed in the above expressions. Therefore, it is essential for n in the circuit of Fig. 2.3 to be greater than 2. This way, the net generated energy will be positive in this harvester.

2.4.3.3   The Net Generated Energy and the Conduction Losses

The energy that is delivered to the load is calculated in (2.72). The expression in (2.23) is used in the following steps to calculate the net generated energy in this harvester.

To find the amount of energy that is obtained from C_r during the investment phase (E_{ri}), the expression in (2.25) should be used. For the harvester of Fig. 2.10, V_{ri} (the voltage across C_r) at the kth cycle is $V_{Cr}(k)$. The integral of the current that goes through C_r during the investment phase is also required for using (2.25). Calculation of this integral is discussed in detail in Sect. 3.1. In this harvester, the voltages across C_r and C_V are $V_{Cr}(k)$ and zero, respectively, at the start of the investment phase. At this moment, S_1 turns on and C_r connects to C_V, when $C_V = C_{max}$. Therefore, the integral of i_{ri} is as follows in this harvester according to Sect. 3.1:

$$\int i_{ri} dt = C_t \Delta V \Rightarrow \int i_{ri} dt = \frac{C_r C_{max}}{C_r + C_{max}} (V_{Cr}(k) - 0),$$

(2.74)

where C_t here is the series equivalent capacitance of C_r and C_{max}.

Replacing V_{ri} with $V_{Cr}(k)$ and the above value for the integral of i_{ri} in (2.25):

$$E_{ri}(k) = \frac{1}{2} C_t V_{Cr}(k) \left(2V_{Cr}(k) - \frac{C_t}{C_r} V_{Cr}(k) \right)$$

$$\Rightarrow E_{ri}(k) = \frac{1}{2} C_t V_{Cr}{}^2(k) \left(2 - \frac{C_t}{C_r} \right).$$

(2.75)

Replacing $V_{Cr}(k)$ with (2.63) and C_t with the series equivalent capacitance of C_r and C_{max} in the above expression:

$$E_{ri}(k) = \frac{1}{2} \frac{C_r C_{max} (2C_r + C_{max})}{(C_r + C_{max})^2} V_r{}^2 \left(\frac{C_r}{C_r + C_{max}} \right)^{2k}.$$

(2.76)

Using the above expression for $E_{ri}(k)$, the total energy that is obtained from C_r is calculated as follows:

$$E_{ri} = \sum_{k=0}^{k=\infty} E_{ri}(k)$$

$$\Rightarrow E_{ri} = \frac{1}{2} \frac{C_r C_{max} (2C_r + C_{max})}{(C_r + C_{max})^2} V_r{}^2 \sum_{k=0}^{k=\infty} \left(\frac{C_r}{C_r + C_{max}} \right)^{2k},$$

(2.77)

where the sum of the geometric series in the above is calculated as follows:

$$\sum_{k=0}^{k=\infty} \left(\frac{C_r}{C_r + C_{max}} \right)^{2k} = \frac{1}{1 - \left(\frac{C_r}{C_r + C_{max}} \right)^2} = \frac{(C_r + C_{max})^2}{C_{max} (2C_r + C_{max})}.$$

(2.78)

Replacing the above value in (2.77):

$$E_{ri} = \frac{1}{2} C_r V_r{}^2.$$

(2.79)

As expected, the total energy that is obtained from C_r until the voltage across it reaches to zero is the same as the initial energy in this capacitor. Replacing the above values for E_{ri} and E_{rr} in (2.23):

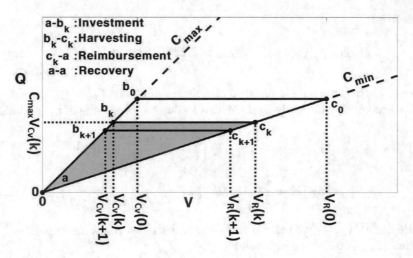

Fig. 2.14 The QV diagram of the circuit in Fig. 2.10

$$E_{net} = E_{rr} - E_{ri}$$

$$\Rightarrow E_{net} = \frac{1}{2}C_r V_r^2 \left(\frac{nC_r - (2C_r + C_{max})}{2C_r + C_{max}} \right). \tag{2.80}$$

According to the above expression, the net generated energy in this harvester is negative if the condition in (2.73) is not satisfied on the value of n. In the case that this condition is not satisfied, the amount of energy that is delivered to the load is less than the amount of energy that is obtained from C_r in each cycle.

The expression in (2.29) is used to find the total conduction losses in this harvester. In this expression, the amount of deliverable energy is required. This energy may be obtained by calculating the enclosed area in the QV diagram of this harvester, according to Sect. 2.2. The QV diagram for this harvester is shown in Fig. 2.14. The phases of operation are specified in this figure for the kth cycle. Each cycle makes an enclosed area in the QV diagram, and the enclosed area is different for each cycle in this harvester. According to the QV diagram, the deliverable energy in the kth cycle is found as follows:

$$E_{del}(k) = \frac{1}{2}C_{max}V_{Cv}(k)\Big(V_R(k) - V_{Cv}(k)\Big)$$

$$\Rightarrow E_{del}(k) = \frac{1}{2}C_{max}(n - 1)V_{Cv}(k)^2, \tag{2.81}$$

where $V_R(k)$ is replaced with $nV_{Cv}(k)$ based on (2.66). Therefore, the total deliverable energy is obtained as follows:

$$E_{del} = \sum_{k=0}^{k=\infty} E_{del}(k)$$

$$\Rightarrow E_{del} = \frac{1}{2} C_{max}(n-1) V_r^2 \sum_{k=0}^{k=\infty} \left(\frac{C_r}{C_r + C_{max}}\right)^{2k+2}$$

$$\Rightarrow E_{del} = \frac{1}{2} C_r V_r^2 \left(\frac{(n-1)C_r}{2C_r + C_{max}}\right), \tag{2.82}$$

where $V_{Cv}(k)$ is replaced with its value in (2.64) and the sum of the geometric series is replaced with its value in (2.71).

Using the expression in (2.29) to find the total conduction losses and the values for E_{net} and E_{del} in (2.80) and (2.82), respectively:

$$E_{loss-tot} = E_{del} - E_{net}$$

$$\Rightarrow E_{loss-tot} = \frac{1}{2} C_r V_r^2 \left(\frac{C_r + C_{max}}{2C_r + C_{max}}\right). \tag{2.83}$$

Considering $C_r \gg C_{max}$ in the above expression, the conduction loss would be half of the initial stored energy in C_r. In this case, $n/2$ times of the initial stored energy in C_r is delivered to the load, based on (2.72). In case that $C_r \ggg C_{max}$, the conduction losses are more than half of the initial stored energy in C_r, and the delivered energy to the load is less than $n/2$ times of this energy.

2.5 Conclusion

Calculating different energies in electrostatic harvesters is explained based on a system-level model in this chapter. A discrete analysis method is also presented, which is helpful in finding closed-form expressions. The presented techniques then are applied to a non-sustainable circuit-level example. Although an amount of energy is generated in non-sustainable harvesters, they stop operating after a while when the charge in their storage component depletes. In the following chapters, sustainable harvesters that have improved performance in this regard are discussed. Fundamentals, naming conventions, and calculation techniques that are presented in this chapter are used in evaluating these harvesters.

References

1. K.H. Rosen, *Discrete Mathematics and Its Applications*, 4th edn. (McGraw-Hill, New York, 1999)

Chapter 3
Switched-Capacitor Electrostatic Harvesters

Abstract This chapter focuses on sustainable switched-capacitor electrostatic harvesters. The variable capacitor in this type of electrostatic harvester connects to the storage component (a battery or a large capacitor) through the investment and reimbursement circuitries. These harvesters use only switches (and no inductor) in their investment and reimbursement circuitries. The chapter begins with fundamentals in analyzing this type of electrostatic harvesters. Then, the net generated energy and conduction losses are calculated for typical harvesters in this category. Using the presented discrete analysis in Chap. 2, the charge-depletion issue is evaluated for the typical switched-capacitor harvesters. Later in this chapter, a harvester with adjustable output voltage is discussed. This harvester is capable of overcoming the charge-depletion issue, owing to the employed switching scheme. A modified version of this harvester is then used in a hybrid energy harvesting system, where the generated energy is used for charging a rechargeable battery. Finally, practical considerations in implementing two energy harvesting systems based on this harvester are discussed. These systems are presented for energy harvesting from knee joint and diaphragm muscle movements.

3.1 Switched-Capacitor Charge Transfer

In this section, the charge transfer between a capacitor and a storage component (a battery or a large capacitor) is discussed, where these components are connected to each other using only active switches (transistors). A transistor connects these components when the voltage across them is different at the moment of connection. The impact of the voltage difference and the switching period on the conduction losses and the current shape are detailed in this section.

© The Author(s), under exclusive license to Springer Nature Switzerland AG 2022
S. H. Daneshvar et al., *Design of Miniaturized Variable-Capacitance Electrostatic Energy Harvesters*, https://doi.org/10.1007/978-3-030-90252-0_3

3.1.1 Using a Transistor

A transistor can connect a capacitor and a storage component at any moment with any voltage difference between these components. This flexibility is widely used in a group of electrostatic harvesters. MOSFET transistors are preferred, since their input power dissipation is negligible compared to BJT transistors. The transistor should operate in triode region when conducting to reduce the conduction losses. In deep-triode region, the transistor may be modelled with a constant resistor. A MOSFET transistor operates in deep-triode region, if the following condition is satisfied [1]:

$$V_{GS} \gg \frac{1}{2} V_{DS} + V_{th}, \tag{3.1}$$

where V_{DS} is the voltage between the Drain and Source terminals of the transistor and V_{th} is the threshold voltage. To satisfy the above condition, the control circuit should generate a Gate–Source pulse for the transistor such that the above relation holds. Selecting a transistor with negligible quiescent current, the transistor may be modelled with an ideal off-switch when it is not conducting. Therefore, a transistor is modelled with an on-resistor in series with an ideal switch in this section. This model is valid if the quiescent current of the transistor is negligible and the transistor operates in deep-triode region when it conducts.

3.1.2 Current Expression

Figure 3.1 shows switched-capacitor charge transfer between C_1 and C_2, when a transistor is used. C_2 is considered as the storage component. Same steps as following may be taken to find the current if the storage component is a battery. In this figure, the model for the transistor is considered to be an on-resistance in series with an ideal switch. Therefore, R_S in this figure is

Fig. 3.1 Switched-capacitor charge transfer between two capacitors

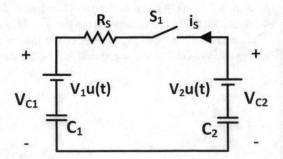

$$R_S = R_{on} + R_{s-c1} + R_{s-c2}, \tag{3.2}$$

where R_{on} is the on-resistance of the transistor, and R_{s-c1} and R_{s-c2} are the equivalent series resistances of C_1 and C_2, respectively. V_1 and V_2 are the initial voltages across C_1 and C_2 before S_1 turns on. The current that goes through C_1, C_2, and R_S is zero before S_1 turns on: $i_S = 0$. At $t = t_s$, S_1 turns on and the charge transfer between these capacitors begins. S_1 turns off at $t = t_e$ and i_S becomes zero again. V_1 and V_2 are declared as unit functions to reflect the effect of S_1 turning on at $t = t_s$. The direction of i_S is specified for the case that $V_2 > V_1$. In this case, an amount of charge transfers from C_2 to C_1 when S_1 turns on. Writing KVL in this circuit, the expression for $i_S(t)$ is obtained as follows:

$$\frac{1}{C_1} \int i_S dt + \frac{1}{C_2} \int i_S dt + R_S i_S = (V_2 - V_1)u(t - t_s)$$

$$\Rightarrow R_S i_S' + \frac{1}{C_t} i_S = (V_2 - V_2)\delta(t - t_s), \tag{3.3}$$

where C_t is the series equivalent capacitance of C_1 and C_2:

$$C_t = \frac{C_1 C_2}{C_1 + C_2}. \tag{3.4}$$

Since initial voltages across C_1 and C_2 are represented as unit functions in the circuit of Fig. 3.1, $i_S(t_s^-) = 0$. Using the above expression:

$$i_S(t_s^+) = \frac{V_2 - V_1}{R_S}, \tag{3.5}$$

therefore,

$$i_S(t) = \frac{\Delta V}{R_S} e^{\frac{-(t-t_s)}{\tau}}, \qquad t_s < t < t_e, \tag{3.6}$$

where

$$\tau = R_S C_t, \qquad \Delta V = V_2 - V_1. \tag{3.7}$$

In the above expression, τ is the time constant of the conducting path between C_1 and C_2 in Fig. 3.1. This time constant depends on the series equivalent capacitance (including C_1 and C_2) and the series equivalent resistance of this path.

In case that the storage component is a battery, C_2 in Fig. 3.1 is replaced with $V_B u(t)$. The battery is represented using a unit function to consider the impact of S_1 turning on at $t = t_s$. The current is the same as in (3.6) where

$$\Delta V = V_B - V_1, \quad \tau = R_S C_1. \tag{3.8}$$

Figure 3.2 shows the impact of changes in the parameters of i_S in (3.6). In Fig. 3.2a, $i_S(t)$ is plotted for different values of ΔV, while the values of R_S and C_t are constant. These values are shown in the figure. As can be seen, higher values of ΔV result in higher currents. The time constant of the conducting path is the same for all the plots in Fig. 3.2a, since R_S and C_t are the same. In this figure, S_1 turns off at $t = t_e$ and t_e is chosen to be five times of the time constant of the conducting path: $t_e = 2\,\mu s$. In Fig. 3.2b and c, the time constant of the conducting path is different for each of the plots, since R_S or C_t changes for each plot. t_e is chosen to be five times of the largest time constant in these figures. The aim of these figures is to illustrate the impact of changing different parameters on the shape of i_S. Therefore, t_e is chosen to be long enough. However, S_1 may turn off at any point and t_e may be any time after t_s. This case will be discussed shortly in the following subsections.

In Fig. 3.2b, $i_S(t)$ is plotted for different values of C_t, while the values of ΔV and R_S are constant. These values are shown in this figure. As can be seen, the higher the value of C_t, the higher the average current is. However, the peak current is the same for all the cases in this figure, since R_S does not change in these plots. Finally, Fig. 3.2c shows the impact of changes in R_S, while ΔV and C_t are constant. Later in this section, it is proved that the integral of i_S is the same for any value of R_S, provided that the switch is on long enough.

3.1.3 Switching Scenarios

In electrostatic switched-capacitor harvesters, the variable capacitor connects to a storage component during the investment and reimbursement phases in a full operating cycle. The net generated energy in the storage component and the total conduction losses may be expressed based on the integral of the currents that go through the storage component during these phases. This is detailed under the second solution for calculating net generated energy in the storage component in Sect. 2.3. Therefore, the integral of $i_S(t)$ is investigated for the following two switching scenarios. Later in this chapter, the net generated energy and the total conduction losses are calculated for typical electrostatic harvesters of this category.

During the time that S_1 is on in the circuit of Fig. 3.1, an amount of energy is obtained from C_2, since $V_2 > V_1$. A part of this energy is transferred to C_1 and the rest is dissipated. The energy flow and the energy losses are discussed in the following Sect. 3.1.4. The amount of transferred and dissipated energies depends on how long S_1 is on. The integral of $i_S(t)$ (that is defined in (3.6)) is calculated as follows:

$$\int_{t_s}^{t_e} i_S(t)dt = C_t \Delta V \left(1 - e^{\frac{-(t_e - t_s)}{\tau}}\right). \tag{3.9}$$

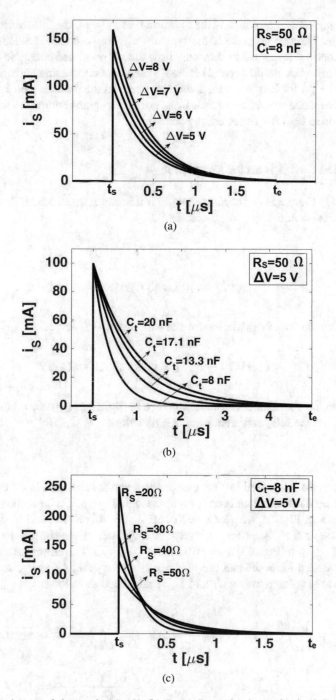

Fig. 3.2 The impact of changes in (**a**) ΔV, (**b**) C_t, (**c**) R_S on the shape of i_S in (3.6)

Depending on $t_e - t_s$, which is the duration that S_1 is on, the following scenarios are possible. The first scenario is the typical switching scenario. Nonetheless, the second scenario is discussed to showcase how the net generated energy in switched-capacitor harvesters would depend on the resistance of conducting paths if a switch does not turn on for long enough. This may occur in a harvester with large series equivalent capacitance and resistance in its conducing paths and/or the harvester is operated with a high frequency energy source.

3.1.3.1 The Switch Is on for Long Enough

The switch in Fig. 3.1 is on for long enough, if it is on for much longer than the time constant of the conducting path; therefore,

$$t_e - t_s \gg \tau \Rightarrow e^{\frac{-(t_e - t_s)}{\tau}} \approx 0$$

$$\Rightarrow \int_{t_s}^{t_e} i_S(t)dt = C_t \Delta V. \tag{3.10}$$

Repeating the above calculation for the case that $t_e - t_s = 5\tau$:

$$\int_{t_s}^{t_e} i_S(t)dt = C_t \Delta V \left(1 - e^{-5}\right) = 0.993\, C_t \Delta V \approx C_t \Delta V. \tag{3.11}$$

Based on the above calculations, the switch in Fig. 3.1 is considered to be on for long enough if the following condition is satisfied:

$$t_e - t_s \geq 5\tau. \tag{3.12}$$

Therefore, the integral of $i_S(t)$ is independent on the resistance in the conducting path, provided that the switch is on for long enough. Figure 3.3 shows the simulation of the circuit in Fig. 3.1 where $C_1 = 10\,\text{nF}$, $C_2 = 40\,\text{nF}$, $V_1 = 1\,\text{V}$, $V_2 = 5\,\text{V}$, $R_S = 100\,\Omega$, and S_1 is on for $4\,\mu\text{s}$ (which is 5 times of the time constant of the conducting path). In Fig. 3.3, $t_s = 10\,\mu\text{s}$ and $t_e = 14\,\mu\text{s}$. The current is almost zero before the switch turns off and the voltages across C_1 and C_2 merge to the same value. The final voltage across C_1 and C_2 is calculated as follows:

$$V_{1f} = V_1 + \frac{1}{C_1} \int_{t_s}^{t_e} i_S(t)dt = V_1 + \frac{C_t}{C_1} \Delta V = \frac{C_1 V_1 + C_2 V_2}{C_1 + C_2} = 4.2\,\text{V}, \tag{3.13}$$

$$V_{2f} = V_2 - \frac{1}{C_2} \int_{t_s}^{t_e} i_S(t)dt = V_2 - \frac{C_t}{C_2} \Delta V = \frac{C_1 V_1 + C_2 V_2}{C_1 + C_2} = 4.2\,\text{V}, \tag{3.14}$$

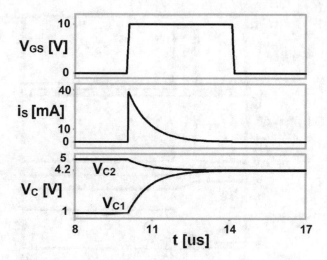

Fig. 3.3 Simulation of the circuit in Fig. 3.1, where $C_1 = 10\,\text{nF}$, $C_2 = 40\,\text{nF}$, $V_1 = 1\,\text{V}$, $V_2 = 5\,\text{V}$, $R_S = 100\,\Omega$, and S_1 is on for 5 times of the time constant of the conducting path

where V_{1f} and V_{2f} are the final voltages across C_1 and C_2 at the end of this charge transfer. These voltages are equal in this case, since the switch is on for a long enough time. Following the charge conservation law, the same expression as above is obtained for V_{1f} and V_{2f}. In the above solution, the integral of $i_S(t)$ needs to be calculated to find V_{1f} and V_{2f}. However, calculating the integral of $i_S(t)$ is the preferred solution throughout this book since knowing this parameter simplifies the calculations of the net generated energy and the total conduction losses. This is detailed in Sect. 2.3.

3.1.3.2 The Switch Is Not on for Long Enough

According to the discussions in the above, the switch in Fig. 3.1 is not on for long enough, if the following condition is met:

$$t_e - t_s < 5\tau. \tag{3.15}$$

In this case, the integral of $i_S(t)$ becomes dependent on the resistance of the conducting path. This is evident from (3.9), where $\tau = R_S C_t$. Figure 3.4 shows the simulation of the circuit in Fig. 3.1 where $C_1 = 10\,\text{nF}$, $C_2 = 40\,\text{nF}$, $V_1 = 1\,\text{V}$, $V_2 = 5\,\text{V}$, $R_S = 100\,\Omega$, and S_1 is on for $0.8\,\mu\text{s}$ (which is equal to the time constant of the conducting path). In Fig. 3.4, $t_s = 10\,\mu\text{s}$ and $t_e = 10.8\,\mu\text{s}$. The current is not zero before the switch turns off and the voltages across C_1 and C_2 do not merge to the same value. The final voltages across C_1 and C_2 are calculated as follows:

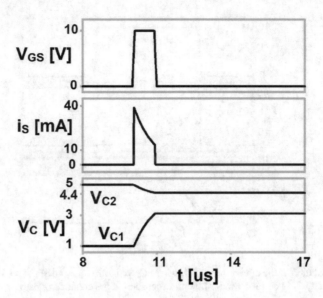

Fig. 3.4 Simulation of the circuit in Fig. 3.1, where $C_1 = 10\,\text{nF}$, $C_2 = 40\,\text{nF}$, $V_1 = 1\,\text{V}$, $V_2 = 5\,\text{V}$, $R_S = 100\,\Omega$, and S_1 is on for a time equal the time constant of the conducting path: $t_e - t_s = \tau = 0.8\,\mu\text{s}$

$$V_{1f} = V_1 + \frac{1}{C_1} \int_{t_s}^{t_e=t_s+\tau} i_S(t)dt = V_1 + \frac{C_t}{C_1} \Delta V \left(1 - e^{-1}\right)$$

$$\Rightarrow V_{1f} = \frac{C_1 V_1 + C_2 V_2}{C_1 + C_2} - \frac{C_2}{C_1 + C_2}(V_2 - V_1)e^{-1} = 4.2 - 1.18 = 3.02 \text{ V},$$

$$(3.16)$$

$$V_{2f} = V_2 - \frac{1}{C_2} \int_{t_s}^{t_e=t_s+\tau} i_S(t)dt = V_2 - \frac{C_t}{C_2} \Delta V \left(1 - e^{-1}\right)$$

$$\Rightarrow V_{2f} = \frac{C_1 V_1 + C_2 V_2}{C_1 + C_2} + \frac{C_1}{C_1 + C_2}(V_2 - V_1)e^{-1} = 4.2 + 0.29 = 4.49 \text{ V},$$

$$(3.17)$$

where V_{1f} and V_{2f} are the final voltages across C_1 and C_2 at the end of this charge transfer. These voltages are not equal since the switch is on for a time equal to time constant of the conducting path: the switch is not on for long enough.

Based on the above expressions, the final voltages across C_1 and C_2 depend on the conducting path resistance. To illustrate this, another simulation with all the same components value as for Fig. 3.4 is shown in Fig. 3.5, where only R_S is changed to $50\,\Omega$. In this figure, $t_s = 10\,\mu\text{s}$ and $t_e = 10.8\,\mu\text{s}$; therefore, the switch is on for the same duration. However, the final voltages across C_1 and C_2 are different

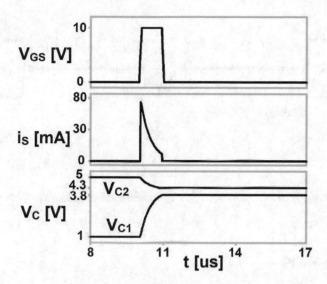

Fig. 3.5 Simulation of the circuit in Fig. 3.1 with all the components value the same as for Fig. 3.4 where S_1 is on for the same duration with the exception of $R_S = 50\,\Omega$

compared to Fig. 3.4. Following the same steps as above, V_1 and V_2 are obtained 3.77 and 4.3 V, respectively. These results are the same as the simulation results in Fig. 3.5.

3.1.4 Energy Transfer and Conduction Losses

Figure 3.6 shows the energy transfer between C_1 and C_2 and the conduction losses in R_S in the circuit of Fig. 3.1. In this figure, an energy expression is shown over each line. The steps to derive the energy expressions are detailed in the following paragraphs. These steps are based on the method explained in Sect. 2.3.2 under the second solution. In this solution, the conduction losses are obtained based on the integral of i_S instead of integral of squared i_S to simplify the energy calculations.

During the charge transfer period, the voltage across C_2 decreases to V_{2f} from V_2. The arrow between V_2 and V_{2f} in Fig. 3.6 is placed downward to reflect the decreasing voltage across C_2. Therefore, an amount of energy is transferred from this capacitor at the end of this period. This energy is calculated in the following steps and is labelled as E_{C2} over the related arrow in Fig. 3.6:

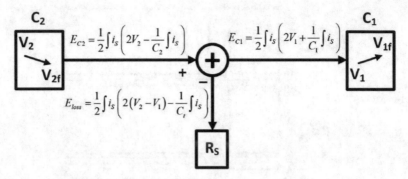

Fig. 3.6 The energy transfer between C_1 and C_2 and the conduction losses in R_S in the circuit of Fig. 3.1

$$V_{2f} = V_2 - \frac{1}{C_2} \int i_S dt,$$

$$E_{C2} = \frac{1}{2} C_2 \left(V_2{}^2 - V_{2f}{}^2 \right) = \frac{1}{2} C_2 \left(V_2{}^2 - \left(V_2 - \frac{1}{C_2} \int i_S dt \right)^2 \right),$$

$$\Rightarrow E_{C2} = \frac{1}{2} \int i_S dt \left(2V_2 - \frac{1}{C_2} \int i_S dt \right). \tag{3.18}$$

However, the voltage across C_1 increases to V_{1f} from V_1 during this phase. The arrow between V_1 and V_{1f} in Fig. 3.6 is placed upward to reflect the increasing voltage across C_1. Therefore, an amount of energy is received by this capacitor at the end of this period. This energy is calculated in the following steps and is labelled as E_{C1} over the related arrow in Fig. 3.6:

$$V_{1f} = V_1 + \frac{1}{C_1} \int i_S dt,$$

$$E_{C1} = \frac{1}{2} C_1 \left(V_{1f}{}^2 - V_1{}^2 \right) = \frac{1}{2} C_1 \left(\left(V_1 + \frac{1}{C_1} \int i_S dt \right)^2 - V_1{}^2 \right),$$

$$\Rightarrow E_{C1} = \frac{1}{2} \int i_S dt \left(2V_1 + \frac{1}{C_1} \int i_S dt \right). \tag{3.19}$$

Due to the conduction losses in R_S, not all the energy that is transferred from C_2 (E_{C2}) is received by C_1 (E_{C1}). The conduction losses in R_S are calculated as follows and are labelled as E_{loss} over the related arrow in Fig. 3.6:

$$E_{loss} = E_{C2} - E_{C1} = \frac{1}{2} \int i_S dt \left(2 \left(V_2 - V_1 \right) - \frac{1}{C_t} \int i_S dt \right), \tag{3.20}$$

where C_t is defined in (3.4). Based on the above expression, the conduction losses dissipated in R_S do not depend on the value of R_S if the integral of i_S is independent of R_S. As explained in Sect. 3.1.3, the integral of i_S does not depend on R_S when the switch is on for $t_{s-on} > 5\tau$ (where τ is the time constant of the conducting path). For this case:

$$\int i_S dt = C_t (V_2 - V_1) \Rightarrow E_{loss} = \frac{1}{2} C_t (V_2 - V_1)^2. \tag{3.21}$$

According to (3.21), the conduction losses between two capacitors with a specific voltage difference $(V_2 - V_1)$ are the same for any R_S in the conducting path between them, when the switch is on for long enough. Interestingly, the same conduction losses occur for the case that the resistance of conducting path is zero. According to (2.20), the conduction losses may also be expressed as follows:

$$E_{loss} = R_S \int i_S^2 dt. \tag{3.22}$$

Writing the conduction losses as above, one may wonder that how the conduction losses are not zero for $R_S = 0$. However, it should be noted that the current tends to infinity when $R_S \to 0$, based on (3.6). Therefore, the above multiplication is not equal to zero when $R_S = 0$ due to an impulse current at $t = t_s$.

3.1.5 Charge Conservation Law

Based on this law, the amount of charge in an isolated system is constant. In the circuit of Fig. 3.1, the sum of the charges in C_1 and C_2 is equal to the following value before S_1 turns on:

$$q_i = C_1 V_1 + C_2 V_2. \tag{3.23}$$

Considering the voltage across these capacitors is equal to V_f after S_1 turns on for long enough, the sum of the charges in C_1 and C_2 is equal to

$$q_f = C_1 V_f + C_2 V_f. \tag{3.24}$$

Therefore, the following relation is true based on the charge conservation law:

$$q_i = q_f \Rightarrow V_f = \frac{C_1 V_1 + C_2 V_2}{C_1 + C_2}. \tag{3.25}$$

Using the above expression is preferred when the intention is finding the final voltage across two capacitors in a scenario similar to above.

In (3.13) and (3.14), the final voltage is found using the integral of the current that goes through the capacitors. This method is preferred when more than two capacitors are connected through a switch and when the calculation of energy is intended.

3.1.6   Overall View

A time constant is defined for a conducting path. The conduction losses do not depend on the resistance of the path, when two capacitors with different voltages connect to each other for a long enough time (more than five times of the time constant of the path) before disconnecting. However, the resistance of the path impacts on the conduction losses, if the capacitors are not connected for long enough. This is true for connecting two capacitors with different voltages through a switch. The case where a variable capacitor connects to a storage component through a diode and the case where an inductor is present between these capacitors are different. These cases are discussed in Chaps. 4 and 5, respectively.

3.2   Study of Elementary Switched-Capacitor Harvesters

The fundamentals of charge transfer between capacitors and storage components using transistors are explained in the previous section. This discussion along with the discrete analysis method that was presented in Chap. 2 is referred to study elementary electrostatic harvesters in this section. The electrostatic harvesters in this section are sustainable and use only storage components (batteries or large capacitors), variable capacitor, and active switches (transistors). The charge-depletion issue that was discussed for non-sustainable harvesters in Chap. 2 is investigated for the elementary sustainable harvesters in this section.

3.2.1   First Elementary Harvester

In this section, an elementary harvester that uses a single active switch between a variable capacitor and a reservoir capacitor is evaluated. The core of the harvester is studied first, and then the cases where a load is placed in series and in parallel with the reservoir capacitor are investigated.

3.2.1.1   The Core Harvester

Figure 3.7 shows the core of an elementary electrostatic harvester with no load. The reservoir capacitor may be replaced with a battery with no impact on the operation of this harvester. Figure 3.8 shows the voltages across C_V and C_r and the current that goes through C_r in this harvester. Based on this figure, the operation of this harvester is explained as follows. Simultaneously, the steps for calculating the voltages across C_V and C_r based on the discrete analysis are followed.

Fig. 3.7 An elementary harvester with no load

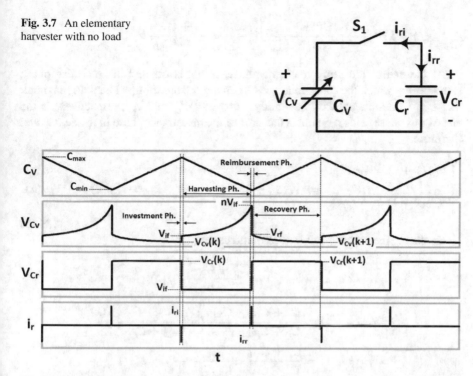

Fig. 3.8 The voltages across C_r and C_V and the current that goes through C_r of the circuit in Fig. 3.7

Phases of operation are shown in Fig. 3.8 for the kth cycle. At the start of the investment phase in this cycle, S_1 is off and $C_V = C_{max}$. At this moment, the voltage across the reservoir capacitor is $V_{Cr}(k)$, and the voltage across the variable capacitor is $V_{Cv}(k)$. The switch turns on and C_V connects to C_r. Provided that the switch remains on for long enough, the voltage across C_V and C_r at the end of the investment phase of the kth cycle is

$$V_{if}(k) = \frac{C_{max} V_{Cv}(k) + C_r V_{Cr}(k)}{C_{max} + C_r}, \qquad (3.26)$$

where the charge conservation law is used to find the above expression. At the end of this phase, S_1 turns off and the harvesting phase starts during which the capacitance of the variable capacitor changes from maximal to minimal. Therefore, the voltages across C_V and C_r are nV_{if} and V_{if}, respectively, at the end of the harvesting phase. At this moment (the start of the reimbursement phase), S_1 turns on and C_V connects to C_r, while $C_V = C_{min}$. Provided that S_1 remains on for long enough, the voltage across C_V and C_r at the end of the reimbursement phase of the kth cycle is equal to

$$V_{rf}(k) = \frac{C_r V_{if}(k) + C_{min} n V_{if}(k)}{C_r + C_{min}} = \frac{C_r + C_{max}}{C_r + C_{min}} V_{if}(k). \tag{3.27}$$

At the end of the reimbursement phase, S_1 turns off and the recovery phase starts during which the capacitance of C_V changes from minimal back to maximal. Therefore, the voltages across C_V and C_r are $(1/n)V_{rf}$ and V_{rf}, respectively, at the end of this phase. This moment is the start of the investment phase in the next cycle; therefore,

$$\begin{cases} V_{Cr}(k+1) = V_{rf}(k) \\ V_{Cv}(k+1) = \dfrac{1}{n} V_{rf}(k) \end{cases} \Rightarrow V_{Cv}(k+1) = \frac{1}{n} V_{Cr}(k+1) \Rightarrow V_{Cv}(k) = \frac{1}{n} V_{Cr}(k). \tag{3.28}$$

Replacing the above result in (3.26) and using (3.27):

$$V_{Cr}(k+1) = V_{rf}(k) = \frac{C_r + C_{max}}{C_r + C_{min}} V_{if}(k)$$

$$\Rightarrow V_{Cr}(k+1) = \frac{C_r + C_{max}}{C_r + C_{min}} \cdot \frac{C_{max} \frac{1}{n} V_{Cr}(k) + C_r V_{Cr}(k)}{C_{max} + C_r}$$

$$\Rightarrow V_{Cr}(k+1) = \frac{C_r + C_{max}}{C_r + C_{min}} \cdot \frac{C_{min} + C_r}{C_{max} + C_r} V_{Cr}(k)$$

$$\Rightarrow V_{Cr}(k+1) = V_{Cr}(k). \tag{3.29}$$

Following the same steps, the below result is obtained for the voltage across C_V:

$$V_{Cv}(k+1) = V_{Cv}(k). \tag{3.30}$$

The above formulas for $V_{Cr}(k)$ and $V_{Cv}(k)$ are valid as long as the relation in (3.28) is true between $V_{Cv}(k)$ and $V_{Cr}(k)$. This relation might not be true before the harvester starts its operation for $k = 0$, since at this moment the voltage across C_V and C_r may be any arbitrary value. Considering that $V_{Cr}(0) = V_r$ and $V_{Cv}(0) = 0$ are the initial voltages across these capacitors, the voltage across C_V and C_r at the end of the first reimbursement phase is obtained as follows based on (3.26) and (3.27):

$$V_{rf}(0) = \frac{C_r + C_{max}}{C_r + C_{min}} \cdot \frac{C_r}{C_r + C_{max}} V_r \Rightarrow V_{rf}(0) = \frac{C_r}{C_r + C_{min}} V_r. \tag{3.31}$$

Using the above expression:

$$V_{Cr}(1) = V_{rf}(0), \tag{3.32}$$

$$V_{Cv}(1) = \frac{1}{n} V_{rf}(0). \tag{3.33}$$

Therefore, the relation in (3.28) between $V_{Cv}(k)$ and $V_{Cr}(k)$ is true for $k \geq 1$ and:

$$V_{Cr}(k) = \begin{cases} V_r & k = 0 \\ \dfrac{C_r}{C_r + C_{min}} V_r & k \geq 1 \end{cases} \qquad V_{Cv}(k) = \begin{cases} 0 & k = 0 \\ \dfrac{1}{n} \cdot \dfrac{C_r}{C_r + C_{min}} V_r & k \geq 1. \end{cases}$$

(3.34)

The voltages across C_r and C_V at the start of the investment phase of any cycle are obtained using the above expressions. As can be seen, these voltages are constant and do not change from cycle to cycle. Therefore, there is no net generated energy in C_r and no energy lost from C_V during the operation of the harvester. The current that goes through C_r during the investment and reimbursement phases (i_{ri} and i_{rr}) passes through S_1 too. Therefore, all the generated energy in this harvester is lost in the on-resistance of S_1.

3.2.1.2 The Harvester with Load in Series

To utilize a part of the lost energy in S_1, a resistive load is placed in series with S_1 in Fig. 3.9. The directions of i_{ri} and i_{rr} are identified in Fig. 3.7, and they are depicted in Fig. 3.8. Knowing $V_{Cr}(k)$, $V_{Cv}(k)$, and $V_{if}(k)$ from above and based on the operation of this harvester:

$$i_{ri}(t) = \frac{V_{Cr}(k) - V_{Cv}(k)}{R_L + R_{S1}} e^{-\frac{t}{\tau_i}} = \frac{n-1}{n(R_L + R_{S1})} \frac{C_r}{C_r + C_{min}} V_r e^{-\frac{t}{\tau_i}}, \qquad (3.35)$$

$$i_{rr}(t) = \frac{(n-1)V_{if}(k)}{R_L + R_{S1}} e^{-\frac{t}{\tau_r}} = \frac{n-1}{R_L + R_{S1}} \frac{C_r}{C_r + C_{max}} V_r e^{-\frac{t}{\tau_r}}, \qquad (3.36)$$

where

$$\tau_i = \frac{C_r C_{max}}{C_r + C_{max}}(R_L + R_{S1}), \qquad \tau_r = \frac{C_r C_{min}}{C_r + C_{min}}(R_L + R_{S1}). \qquad (3.37)$$

Fig. 3.9 The elementary harvester in Fig. 3.7 with a load in series with the reservoir capacitor

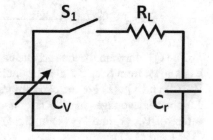

In this harvester, the net generated energy is equal to the delivered energy to the load (R_L) during the investment and reimbursement phases. This is due to the fact that the energy in C_r and C_V does not change during the operation of this harvester; therefore,

$$E_{net} = R_L \int i_{ri}{}^2(t)dt + R_L \int i_{rr}{}^2(t)dt$$

$$\Rightarrow E_{net} = R_L \left(\int i_{ri}{}^2(t)dt + \int i_{rr}{}^2(t)dt \right). \tag{3.38}$$

The total conduction losses are equal to the energy that is lost in R_{S1} during the investment and reimbursement phases, and therefore,

$$E_{loss-tot} = E_{li} + E_{lr} = R_{S1} \int i_{ri}{}^2(t)dt + R_{S1} \int i_{rr}{}^2(t)dt$$

$$\Rightarrow E_{loss-tot} = R_{S1} \left(\int i_{ri}{}^2(t)dt + \int i_{rr}{}^2(t)dt \right). \tag{3.39}$$

The common term in the above expressions is obtained as below, using (3.35) and (3.36):

$$\int i_{ri}{}^2(t)dt + \int i_{rr}{}^2(t)dt = \frac{1}{2} C_r V_r{}^2 \frac{(n-1)^2 C_r{}^2}{R_L + R_S} \frac{C_{max}\left((n+1)C_r + 2C_{max}\right)}{(C_r + C_{max})^2 (nC_r + C_{max})^2}. \tag{3.40}$$

Using the above value in (3.38) and (3.39), E_{net} and $E_{loss-tot}$ are obtained for this harvester. Using (2.29), (3.38), and (3.39), the deliverable energy in this harvester is obtained as follows:

$$E_{loss-tot} = E_{del} - E_{net} \Rightarrow E_{del} = E_{net} + E_{loss-tot}$$

$$\Rightarrow E_{del} = (R_L - R_{S1}) \left(\int i_{ri}{}^2(t)dt + \int i_{rr}{}^2(t)dt \right)$$

$$\Rightarrow E_{del} = \frac{1}{2} C_r V_r{}^2 \frac{(n-1)^2 C_r{}^2 C_{max}\left((n+1)C_r + 2C_{max}\right)}{(C_r + C_{max})^2 (nC_r + C_{max})^2}. \tag{3.41}$$

The QV diagram for the variable capacitor in this harvester is shown in Fig. 3.10. It is known from Sect. 2.2 that the enclosed area in this figure is equal to E_{del}. Since $V_{Cv}(k)$ and $V_{Cr}(k)$ are the constant values in this harvester, $V_{if}(k)$ and $V_{rf}(k)$ are the constant voltages and do not change from a cycle to the next one. Using the values of $V_{Cv}(k)$ and $V_{Cr}(k)$ in (3.34) and the expressions for $V_{if}(k)$ and $V_{rf}(k)$ in (3.26) and (3.27):

Fig. 3.10 The QV diagram of the variable capacitor in the harvester of Fig. 3.7

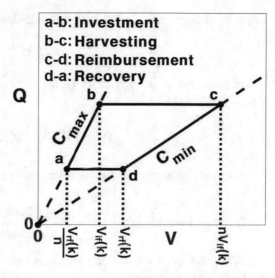

Fig. 3.11 The elementary harvester in Fig. 3.7 with a load in series with the reservoir capacitor

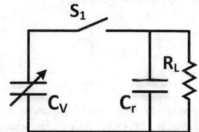

$$V_{if}(k) = \frac{C_r}{C_r + C_{max}} V_r, \qquad V_{rf}(k) = \frac{C_r}{C_r + C_{min}} V_r. \qquad (3.42)$$

Replacing these values in the formula for the enclosed area of the QV diagram in Fig. 3.10, it is confirmed that this area is equal to E_{del} in (3.41).

3.2.1.3 The Harvester with Load in Parallel

Figure 3.11 shows the harvester in Fig. 3.7 with a load in parallel with C_r. Considering the voltages across C_r and C_V are $V_{Cr}(k)$ and $V_{Cv}(k)$ at the kth cycle, $V_{if}(k)$ is obtained the same as in (3.26). During the harvesting phase, C_V is not connected to R_L, since S_1 is off. However, C_r is in parallel with R_L during this phase. Therefore, the voltages across these capacitors at the start of the reimbursement phase ($V_{cr-rs}(k)$ and $V_{cv-rs}(k)$) are as follows:

$$V_{cr-rs}(k) = V_{if}(k)e^{-\frac{T_V}{2\tau_L}}, \qquad V_{cv-rs}(k) = nV_{if}(k), \qquad (3.43)$$

where T_V is the period of a full cycle and is defined in (2.1) and

$$\tau_L = C_r R_L. \tag{3.44}$$

Following the same steps as for (3.28), the below relation is obtained between $V_{Cr}(k)$ and $V_{Cv}(k)$ for this case:

$$V_{Cv}(k) = e^{\frac{T_V}{2\tau_L}} \frac{1}{n} V_{Cr}(k). \tag{3.45}$$

Using the above relation, and the same steps as for (3.29):

$$V_{Cr}(k+1) = \alpha V_{Cr}(k) \qquad \alpha = \frac{C_r e^{-\frac{T_V}{2\tau_L}} + C_{max}}{C_r + C_{max}} \frac{C_r e^{-\frac{T_V}{2\tau_L}} + C_{min}}{C_r + C_{min}} V_{Cr}(k). \tag{3.46}$$

Solving the above relation and assuming that initial voltages across C_r and C_V are V_r and 0, respectively:

$$V_{Cr}(k) = \begin{cases} V_r & k = 0 \\ \dfrac{C_r e^{-\frac{T_V}{2\tau_L}} + C_{max}}{C_r + C_{max}} \dfrac{C_r}{C_r + C_{min}} \alpha^{k-1} & k \geq 1 \end{cases}, \tag{3.47}$$

where α is defined in (3.46) and $\alpha < 1$. Therefore, the voltage across C_r eventually depletes and the harvester stops working.

3.2.1.4 Overall

Evaluating this harvester with discrete analysis in the above, it is revealed that the harvester is not sustainable when a load is placed in parallel with C_r. This is due to the fact that the voltage across this capacitor depletes regardless of the resistance of the load in this case. The load may be placed in series with C_r in this harvester. In this case, the voltages across C_r and C_V do not increase from a cycle to the next cycle and an amount of energy delivers to the load. However, an equivalent parallel resistor exists internally in any capacitor. The value of this resistor depends on the dielectric material and the distance between layers forming the capacitor. Therefore, the voltage across C_r eventually depletes because of this resistor even if the load is placed in series with C_r. In the above analysis, it is shown that charge depletion occurs no matter how large this resistor might be. Overall, this harvester is not sustainable in practice.

3.2.2 Second Elementary Harvester

In this section, another elementary harvester that uses two reservoir capacitors is evaluated. The core of this harvester is studied first. Then, an analysis is performed to investigate the charge-depletion issue in this harvester by considering large equivalent parallel resistances to the two reservoir capacitors.

3.2.2.1 The Core Harvester

Figure 3.12 shows the core of an elementary harvester with two reservoir capacitors. This harvester is presented in [2]. The voltages across these reservoir capacitors and the variable capacitor are plotted in Fig. 3.13. The operation of this harvester is

Fig. 3.12 The elementary harvester with two reservoir capacitors in [2]

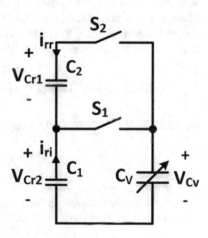

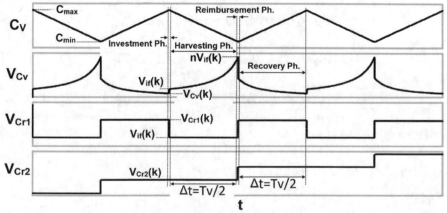

Fig. 3.13 The voltages across C_{r1}, C_{r2}, and C_V of the circuit in Fig. 3.12

explained as follows referring to this figure. Simultaneously, the expressions for the voltages across the capacitors of this harvester are obtained.

At the start of the investment phase in the kth cycle, the voltages across C_{r1}, C_{r2}, and C_V are $V_{Cr1}(k)$, $V_{Cr2}(k)$, and $V_{Cv}(k)$, respectively. These voltages are marked in Fig. 3.13 in the kth cycle. At this moment, S_1 turns on and C_V connects to C_{r1}, while $C_V = C_{max}$. Therefore, the voltage across these capacitors at the end of this phase is

$$V_{if}(k) = \frac{C_{max}V_{Cv}(k) + C_{r1}V_{Cr1}(k)}{C_{max} + C_{r1}}. \tag{3.48}$$

At the end of this phase, S_1 turns off and the harvesting phase starts. The voltages across C_{r1} and C_{r2} do not change during this phase, while the voltage across C_V reaches to $nV_{if}(k)$ at the end of this phase ($n = C_{max}/C_{min}$). Therefore, the voltages across C_{r1} and C_{r2} are $V_{if}(k)$ and $V_{Cr2}(k)$, respectively, at the end of this phase. At this moment, S_2 turns on and C_V connects to C_{r1} and C_{r2}. This starts the reimbursement phase. The integral of the current that goes through these capacitors during the reimbursement phase of the kth cycle is calculated as follows:

$$\int i_{rr}dt(k) = C_t \Delta V(k) = C_t\Big((n-1)V_{if}(k) - V_{Cr2}(k)\Big), \tag{3.49}$$

where C_t is the series equivalent capacitance of C_{r1}, C_{r2}, and C_{min}, and therefore,

$$C_{12} = \frac{C_{r1}C_{r2}}{C_{r1} + C_{r2}}, \qquad C_t = \frac{C_{12}C_{min}}{C_{12} + C_{min}}. \tag{3.50}$$

Therefore, the voltages across C_{r1}, C_{r2}, and C_V at the end of the reimbursement phase ($V_{Cr1-rf}(k)$, $V_{Cr2-rf}(k)$, and $V_{Cv-rf}(k)$) are calculated as follows:

$$V_{Cr1-rf}(k) = V_{if}(k) + \frac{1}{C_{r1}}\int i_{rr}dt(k),$$

$$V_{Cr2-rf}(k) = V_{Cr2}(k) + \frac{1}{C_{r2}}\int i_{rr}dt(k),$$

$$V_{Cv-rf}(k) = nV_{if}(k) - \frac{1}{C_{min}}\int i_{rr}dt(k). \tag{3.51}$$

Using (3.49) and (3.51):

$$V_{Cr1-rf}(k) = \left(1 + \frac{C_t}{C_{r1}}(n-1)\right)V_{if}(k) - \frac{C_t}{C_{r1}}V_{Cr2}(k),$$

$$V_{Cr2-rf}(k) = \frac{C_t}{C_{r2}}(n-1)V_{if}(k) + \left(1 - \frac{C_t}{C_{r2}}\right)V_{Cr2}(k),$$

$$V_{Cv-rf}(k) = \left(n - \frac{C_t}{C_{min}}(n-1)\right) V_{if}(k) + \frac{C_t}{C_{min}} V_{Cr2}(k). \tag{3.52}$$

The recovery phase starts at the end of the reimbursement phase. The voltage across the variable capacitor decreases n times during this phase, while the voltages across C_{r1} and C_{r2} remain constant. At the end of the recovery phase, the investment phase of the $(k+1)$th cycle starts and C_V connects to C_{r1} again. Therefore,

$$V_{if}(k+1) = \frac{C_{r1} V_{Cr1-rf}(k) + C_{max} \frac{V_{Cv-rf}(k)}{n}}{C_{r1} + C_{max}}. \tag{3.53}$$

Using the values of $V_{Cr1-rf}(k)$ and $V_{Cv-rf}(k)$ from (3.52) in the above expression, the following result is obtained:

$$V_{if}(k+1) = V_{if}(k). \tag{3.54}$$

Therefore, $V_{if}(k)$ is constant from one cycle to the next one. Considering the initial conditions:

$$V_{if}(k) = V_{if}(0) = \frac{C_{r1} V_{Cr1}(0) + C_{max} V_{Cv}(0)}{C_{r1} + C_{max}}, \tag{3.55}$$

where $V_{Cr1}(0)$ and $V_{Cv}(0)$ are the initial voltages across C_{r1} and C_V before the harvester starts operating. As can be seen, $V_{if}(k)$ does not depend on k and is constant during the operation of the harvester. To reflect this, $V_{if}(k)$ is replaced with $V_{if}(0)$ in the rest of the analysis of this harvester.

Since the voltage across C_{r2} does not change during the recovery phase, the voltage across C_{r2} at the beginning of the investment phase of $(k+1)$th cycle is

$$V_{Cr2}(k+1) = V_{Cr2-rf}(k)$$

$$\Rightarrow V_{Cr2}(k+1) = \frac{C_t}{C_{r2}}(n-1) V_{if}(0) + \left(1 - \frac{C_t}{C_{r2}}\right) V_{Cr2}(k)$$

$$\Rightarrow V_{Cr2}(k+1) - \left(1 - \frac{C_t}{C_{r2}}\right) V_{Cr2}(k) = \frac{C_t}{C_{r2}}(n-1) V_{if}(0). \tag{3.56}$$

The above expression is a linear non-homogeneous recursive expression. This expression is solved following the steps in Chap. 2 as follows:

$$V_{Cr2}(k) = V_{Cr2-g}(k) + V_{Cr2-p}(k)$$

$$\Rightarrow V_{Cr2}(k) = A\left(1 - \frac{C_t}{C_{r2}}\right)^k + \frac{\frac{C_t}{C_{r2}}(n-1) V_{if}(0)}{1 - \left(1 - \frac{C_t}{C_{r2}}\right)}$$

$$\Rightarrow V_{Cr2}(k) = A\left(1 - \frac{C_t}{C_{r2}}\right)^k + (n-1)V_{if}(0), \tag{3.57}$$

where considering the initial voltage across C_{r2} is $V_{Cr2}(0)$ before the harvester starts operating in the above expression:

$$V_{Cr2}(k) = \left(V_{Cr2}(0) - (n-1)V_{if}(0)\right)\left(1 - \frac{C_t}{C_{r2}}\right)^k + (n-1)V_{if}(0), \quad k \geq 0. \tag{3.58}$$

The voltage across C_{r1} does not change during the recovery phase, and therefore,

$$V_{Cr1}(k+1) = V_{Cr1-rf}(k) = \left(1 + \frac{C_t}{C_{r1}}(n-1)\right)V_{if}(0) - \frac{C_t}{C_{r1}}V_{Cr2}(k), \tag{3.59}$$

where the value of $V_{Cr2}(k)$ is expressed in (3.58). Therefore,

$$V_{Cr1}(k+1) = -\frac{C_t}{C_1}\left(V_{Cr2}(0) - (n-1)V_{if}(0)\right)\left(1 - \frac{C_t}{C_{r2}}\right)^k + V_{if}(0). \tag{3.60}$$

Using the above expression, $V_{Cr1}(k)$ is obtained as follows. This is obtained by replacing k with $k-1$ in the above expression:

$$V_{Cr1}(k) = \begin{cases} V_{Cr1}(0) & k = 0 \\ -\dfrac{C_t}{C_1}\left(V_{Cr2}(0) - (n-1)V_{if}(0)\right)\left(1 - \dfrac{C_t}{C_{r2}}\right)^{k-1} + V_{if}(0) & k \geq 1 \end{cases}, \tag{3.61}$$

where $V_{Cr1}(0)$ is the initial voltage across C_{r1} before the harvester starts operating.

The voltage across C_V is n times less than $V_{Cv-rf}(k)$ in (3.52) at the beginning of the investment phase of $(k+1)$th cycle. This is due to the fact that C_V is isolated and its capacitance changes from minimal to maximal during the recovery phase. Following the same steps as for $V_{Cr1}(k)$ in the above, $V_{Cv}(k)$ is obtained as below:

$$V_{Cv}(k) = \begin{cases} V_{Cv}(0) & k = 0 \\ \dfrac{C_t}{C_{max}}\left(V_{Cr2}(0) - (n-1)V_{if}(0)\right)\left(1 - \dfrac{C_t}{C_{r2}}\right)^{k-1} + V_{if}(0) & k \geq 1 \end{cases}, \tag{3.62}$$

where $V_{Cv}(0)$ is the initial voltage across C_V before the harvester starts operating.

The term that is powered to k or $k-1$ in the expressions for the voltages across C_{r1}, C_{r2}, and C_V is as follows. This term is always less than 1, since the equivalent capacitance of a capacitor in series with other capacitors is always less than the capacitance of that capacitor.

$$\left(1 - \frac{C_t}{C_{r2}}\right) < 1. \tag{3.63}$$

According to the above relation, the voltages across the capacitors in this harvester converge to a final value as below:

$$\lim_{k \to \infty} V_{Cr1}(k) = V_{if}(0),$$

$$\lim_{k \to \infty} V_{Cr2}(k) = (n-1)V_{if}(0),$$

$$\lim_{k \to \infty} V_{Cv}(k) = V_{if}(0), \tag{3.64}$$

where $V_{Cr2}(k)$, $V_{Cr1}(k)$, and $V_{Cv}(k)$ are expressed in (3.58), (3.61), and (3.62), respectively. The operation of this harvester is shown in a longer period in Fig. 3.14. The voltages across C_{r1} and C_{r2} are marked at the kth and $(k+1)$th cycles in this figure. As can be seen, the voltage across C_{r2} increases from a cycle to the next one, and it finally converges to $(n-1)V_{if}(0)$. However, the voltage across C_{r1} decreases from a cycle to the next one, and it finally converges to $V_{if}(0)$.

The QV diagram of this harvester is depicted in Fig. 3.15. The voltage across C_V is $V_{Cv}(0)$ before the harvester starts operating: a_0 marks this moment. In the first cycle, the variable capacitor goes through the $a_0b_0c_0d_0a_1$ path: the enclosed area of this path is equal to the deliverable energy in this harvester in the first cycle, $E_{del}(0)$. The next cycle will have a smaller enclosed area that means that the deliverable energy in the second cycle is less than the deliverable energy in the first cycle. The path that the variable capacitor goes through in the kth cycle is marked with $a_kb_0c_0d_ka_{k+1}$, and the enclosed area of this path is greyed in this figure. Calculating this area, the deliverable energy in the kth cycle is

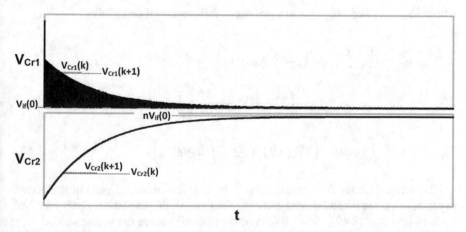

Fig. 3.14 The voltages across C_{r1} and C_{r2} of the circuit in Fig. 3.12, when the harvester operates for a long time

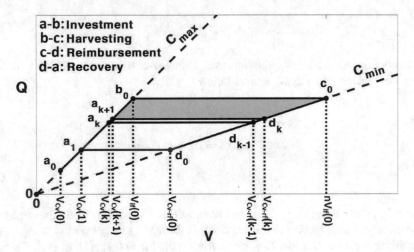

Fig. 3.15 The QV diagram of the circuit in Fig. 3.12

$$E_{del}(k) = \frac{1}{2}C_{max}(n-1)\left(V_{if}^{2}(0) - V_{Cv}^{2}(k+1)\right). \tag{3.65}$$

As can be seen in Fig. 3.15, the deliverable energy decreases from a cycle to the next one. According to the above expression, $E_{del}(k)$ converges to zero since $V_{Cv}(k+1)$ converges to $V_{if}(0)$ based on (3.64).

The net generated energy is as below, since there are two reservoir capacitors in this harvester:

$$E_{net} = E_{net-Cr1} + E_{net-Cr2}, \tag{3.66}$$

where $E_{net-Cr1}$ and $E_{net-Cr2}$ are written as follows based on (2.28):

$$E_{net-Cr1} = \frac{1}{2}\left(\int i_{rr}dt(k) - \int i_{ri}dt(k)\right)$$

$$\left(2V_{Cr1}(k) + \frac{1}{C_{r1}}\left(\int i_{rr}dt(k) - \int i_{ri}dt(k)\right)\right),$$

$$E_{net-Cr2} = \frac{1}{2}\int i_{rr}dt(k)\left(2V_{Cr2}(k) + \frac{1}{C_{r2}}\int i_{rr}dt(k)\right). \tag{3.67}$$

The integral of the investment current in C_{r2} is considered zero in the above expressions, since no investment current goes through this capacitor. The integral of i_{rr} is obtained in (3.49). This value is always positive since the voltage across C_{r2} converges to $(n-1)V_{if}(0)$. Therefore, the net generated energy in C_{r2} is positive and converges to zero. The net generated energy in C_{r1} depends on the difference between the integral of i_{rr} and the integral of i_{ri}. The integral of i_{ri} at the kth cycle is as below:

$$\int i_{ri} dt(k) = C_T \Big(V_{Cr1}(k) - V_{Cv}(k) \Big), \qquad C_T = \frac{C_{r1} C_{max}}{C_{r1} + C_{max}}. \qquad (3.68)$$

Replacing the values of $V_{Cr1}(k)$ and $V_{Cv}(k)$ in the above expression:

$$\int i_{ri} dt(k) = C_t \Big((n-1) V_{if}(0) - V_{Cr2}(0) \Big) \Big(1 - \frac{C_t}{C_{r2}} \Big)^{k-1}. \qquad (3.69)$$

Replacing the value of $V_{Cr2}(k)$ in (3.49) and using the above expression for the integral of i_{ri}, the difference between the integral of i_{rr} and the integral of i_{ri} at the kth cycle is

$$\int i_{rr} dt(k) - \int i_{ri} dt(k) = -\frac{C_t^2}{C_{r2}} \Big((n-1) V_{if}(0) - V_{Cr2}(0) \Big) \Big(1 - \frac{C_t}{C_{r2}} \Big)^{k-1}. \qquad (3.70)$$

As can be seen, this difference is always negative, since $V_{Cr2}(0)$ is less than the final voltage across C_{r2}. Therefore, the net generated energy in C_{r1} is negative and converges to zero. This energy is obtained by replacing $V_{Cr1}(k)$ and the above value in (3.67).

Knowing the values of $E_{del}(k)$ and $E_{net}(k)$ from the above, the total conduction losses are obtained as follows based on (2.29):

$$E_{loss-tot}(k) = E_{del}(k) - E_{net}(k) = E_{del}(k) - E_{net-Cr1}(k) - E_{net-Cr2}(k). \qquad (3.71)$$

Figure 3.16 shows the plot of E_{del}, E_{net1}, E_{net2}, and $E_{loss-tot}$ versus the number of operating cycles, where $C_{min} = 1$ nF, $C_{r1} = C_{r2} = 40$ nF, and $n = 5$. In this figure, the net generated energy in C_{r1} is less than zero and the net generated energy in C_{r2} is maximal for $k = 15$. Eventually, the voltages across C_{r1} and C_{r2} reach to their final value, and E_{del}, E_{net2}, E_{net1}, and $E_{loss-tot}$ converge to zero.

3.2.2.2 Harvester in Practice

In practice, a large resistance exists in parallel with any capacitor. This internal resistor represents the resistance between the two plates of a capacitor. The value of this resistor depends on the dielectric material and the distance between the two plates forming the capacitor. Figure 3.17 shows the harvester in Fig. 3.12 with the inherent parallel resistors of C_{r1} and C_{r2}. Nothing changes regarding the operation of the harvester; therefore, the voltages across the capacitors are the same as depicted in Fig. 3.13. However, the voltages across C_{r1} and C_{r2} decrease during both harvesting and recovery phases. The duration of each of these phases is as long as $T_V/2$ as shown in Fig. 3.13.

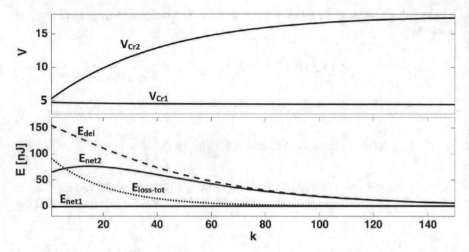

Fig. 3.16 The deliverable energy, net generated energy, and total conduction losses versus the number of operating cycles for the circuit in Fig. 3.12

Fig. 3.17 The elementary harvester in Fig. 3.12 with the inherent parallel resistors of C_{r1} and C_{r2}

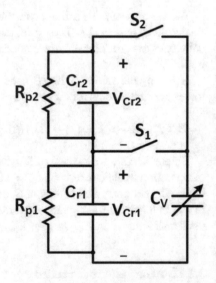

The voltages across C_{r1} and C_V are the same as $V_{if}(k)$ expressed in (3.48) at the end of the investment phase at the kth cycle. At this moment, the voltage across C_{r2} is $V_{Cr2}(k)$. At the end of the harvesting phase (and the start of the reimbursement phase), the voltage across C_V is $nV_{if}(k)$; however, the voltages across C_{r1} ($V_{Cr1-rs}(k)$) and C_{r2} ($V_{Cr2-rs}(k)$) are as below:

$$V_{Cr1-rs}(k) = \alpha V_{Cr1}(k), \qquad V_{Cr2-rs}(k) = \beta V_{Cr2}(k), \qquad (3.72)$$

where:

$$\alpha = e^{-\frac{T_V}{2\tau_{p1}}} < 1, \qquad \beta = e^{-\frac{T_V}{2\tau_{p2}}} < 1, \tag{3.73}$$

and

$$\tau_{p1} = C_{r1}R_{p1}, \qquad \tau_{p2} = C_{r2}R_{p2}. \tag{3.74}$$

At this moment, S_2 turns on and C_V connects to C_{r1} and C_{r2}; therefore, the integral of the current that goes through these capacitors is as follows, based on (3.11):

$$\int i_{rr}dt(k) = C_t\Delta V(k) = C_t\Big((n-\alpha)V_{if}(k) - \beta V_{Cr2}(k)\Big), \tag{3.75}$$

where C_t is defined in (3.50). Therefore, the voltages across these capacitors at the end of the reimbursement phase are

$$V_{Cr1-rf}(k) = \left(\alpha + \frac{C_t}{C_{r1}}(n-\alpha)\right)V_{if}(k) - \beta\frac{C_t}{C_{r1}}V_{Cr2}(k),$$

$$V_{Cr2-rf}(k) = \frac{C_t}{C_{r2}}(n-\alpha)V_{if}(k) + \beta\left(1 - \frac{C_t}{C_{r2}}\right)V_{Cr2}(k),$$

$$V_{Cv-rf}(k) = \left(n - \frac{C_t}{C_{min}}(n-\alpha)\right)V_{if}(k) + \beta\frac{C_t}{C_{min}}V_{Cr2}(k). \tag{3.76}$$

The recovery phase starts, and the voltages across these capacitors at the end of the recovery phase (and the start of the investment phase of $(k+1)$th cycle) are obtained as follows:

$$V_{Cr1-is}(k+1) = \alpha V_{Cr1-rf}(k),$$

$$V_{Cr2-is}(k+1) = \beta V_{Cr2-rf}(k),$$

$$V_{Cv-is}(k+1) = \frac{1}{n}V_{Cv-rf}(k). \tag{3.77}$$

Using the above expressions for the voltages across C_{r1} and C_V at this moment, the voltage across these capacitors at the end of the investment phase of the $(k+1)$th cycle is

$$V_{if}(k+1) = AV_{if}(k) + BV_{Cr2}(k), \tag{3.78}$$

where:

$$A = \frac{\alpha^2 C_{r1} + \frac{C_{max}\left(\frac{\alpha(n-\alpha+1)}{n}C_{12}+C_{min}\right)}{C_{12}+C_{min}}}{C_{r1} + C_{max}}, \qquad B = \frac{\beta(1-\alpha)C_t}{C_{r1}+C_{max}}. \tag{3.79}$$

The voltages across the capacitors of this harvester are expressed at the end of the reimbursement phase of the kth cycle in (3.52) with no resistor in parallel with C_{r1} and C_{r2}. These values are expressed in (3.76) for the case that the inherent parallel resistors are considered for C_{r1} and C_{r2}. Comparing these two cases, it is evident that these voltages are less at the end of this phase when R_{p1} and R_{p2} exist, since $\alpha, \beta < 1$ based on (3.73). The voltages across C_{r1} and C_{r2} are multiplied by α and β once again during the recovery phase. Therefore with R_{p1} and R_{p2}, $V_{Cr1}(k + 1)$ and $V_{Cr2}(k + 1)$ are less than their values in (3.59) and (3.56). To investigate the case, where the impact of R_{p1} and R_{p2} is minimized in decreasing the values of $V_{Cr1}(k + 1)$ and $V_{Cr2}(k + 1)$, the following assumption should be considered:

$$C_{r1}R_{p1} \gg \frac{T_V}{2} \Rightarrow \alpha \approx 1, \qquad C_{r2}R_{p2} \gg \frac{T_V}{2} \Rightarrow \beta \approx 1. \tag{3.80}$$

With the above assumption, the values of A and B in (3.79) are rewritten as follows:

$$A \approx \alpha^2, \qquad B \approx 0. \tag{3.81}$$

Therefore,

$$V_{if}(k + 1) - \alpha^2 V_{if}(k) \approx 0. \tag{3.82}$$

Solving the above linear, homogeneous recursive formula and considering the initial voltages across C_{r1} and C_V ($V_{Cr1}(0)$ and $V_{Cv}(0)$):

$$V_{if}(k) \approx V_{if}(0)\alpha^{2k}, \qquad V_{if}(0) = \frac{C_{r1}V_{Cr1}(0) + C_{max}V_{Cv}(0)}{C_{r1} + C_{max}}. \tag{3.83}$$

Following the steps in the above, the voltage across C_{r2} at the end of the investment phase of the $(k + 1)$th is

$$V_{Cr2}(k + 1) = \beta \frac{C_t}{C_{r2}}(n - \alpha)V_{if}(k) + \beta^2 \left(1 - \frac{C_t}{C_{r2}}\right) V_{Cr2}(k). \tag{3.84}$$

To solve the above linear non-homogeneous recursive formula, the general response should be added to a particular response. Replacing the value of $V_{if}(k)$ from (3.83) in the above expression, a particular response is considered as follows:

$$V_{Cr2-p}(k) = a\alpha^{2k} + b, \tag{3.85}$$

where a and b are the constant coefficients. Replacing the above particular response in (3.84), these coefficients are obtained as below:

$$a = \frac{\beta \frac{C_t}{C_{r2}}(n - \alpha)V_{if}(0)}{\alpha^2 - \beta^2 \left(1 - \frac{C_t}{C_{r2}}\right)}, \qquad b = 0. \tag{3.86}$$

Therefore, a particular response of $V_{Cr2}(k)$ is

$$V_{Cr2-p}(k) = \frac{\beta \frac{C_t}{C_{r2}}(n - \alpha)V_{if}(0)}{\alpha^2 - \beta^2 \left(1 - \frac{C_t}{C_{r2}}\right)}\alpha^{2k}. \tag{3.87}$$

The general response of $V_{Cr2}(k)$ in (3.84) is

$$V_{Cr2-g}(k) = D\left(\beta^2 \left(1 - \frac{C_t}{C_{r2}}\right)\right)^k. \tag{3.88}$$

Therefore, $V_{Cr2}(k)$ is obtained as follows according to the above:

$$V_{Cr2}(k) = V_{Cr2-p}(k) + V_{Cr2-g}(k)$$

$$\Rightarrow V_{Cr2}(k) = \frac{\beta \frac{C_t}{C_{r2}}(n - \alpha)V_{if}(0)}{\alpha^2 - \beta^2 \left(1 - \frac{C_t}{C_{r2}}\right)}\alpha^{2k} + D\left(1 - \frac{C_t}{C_{r2}}\right)^k \beta^{2k}, \tag{3.89}$$

where D in the above is obtained as follows by replacing $k = 0$ in the above and considering the initial voltage across C_{r2} is $V_{Cr2}(0)$:

$$D = V_{Cr2}(0) - \frac{\beta \frac{C_t}{C_{r2}}(n - \alpha)V_{if}(0)}{\alpha^2 - \beta^2 \left(1 - \frac{C_t}{C_{r2}}\right)}. \tag{3.90}$$

3.2.2.3 Overall

According to the above expressions, $V_{if}(k)$ and $V_{Cr2}(k)$ eventually become zero, considering that all the terms in these expressions that are powered to k are less than 1. Therefore, the charge in the capacitors of this harvester eventually depletes, when inherent parallel resistors are considered for the reservoir capacitors. The above analysis shows that the charge depletion occurs in this harvester, regardless of how big these resistors are. Although the generated energy in this harvester is transferred back to the storage components, but the structure still suffers from charge depletion.

Compared to the harvester in Sect. 3.2.1, this harvester is capable of generating a positive net generated energy in one of its storage component. However, the net generated energy in the other storage component of this harvester is negative. This issue eventually causes charge depletion occur in both of the storage components, when the inherent parallel resistors are considered for these storage components. Overall, this harvester is not sustainable in practice.

3.2.3 Charge-Depletion Issue

The sustainable structure is explained in Chap. 2. Both elementary harvesters in this section follow this structure. The generated energy in the variable capacitor of these harvesters is transferred back to their storage components. However, following this structure is not enough to make a harvester sustainable. The second condition for a harvester to be sustainable is that the voltages across its storage components do not deplete eventually in case a resistor is placed in parallel with them. Equivalently, the voltages across the storage components in sustainable harvesters increase from a cycle to another cycle, unless they are capped with a large capacitor or a battery.

3.3 Study of Sustainable Switched-Capacitor Harvesters

In this section, a sustainable switched-capacitor harvester is studied. In the first step, the core of this harvester is discussed and the operation of the harvester is explained. Later, an adjustable output voltage harvester based on this sustainable core harvester is studied. Finally, a structure using this core to charge a battery is detailed.

3.3.1 The Core Harvester

Figure 3.18 shows the sustainable switched-capacitor harvester that is proposed in [3]. The operation of this harvester and the voltages across its capacitors are illustrated in Fig. 3.19. The kth full operating cycle consisting of two half cycles is shown in this figure. The time period for each of the half cycles is equal to T_V (which is defined in (2.1)). The sequence of events in the first half cycle is the same as the harvester in Sect. 3.2.2. C_{r1} charges C_V through S_1 and S_3 during the investment phase of the first half cycle. Later, the generated energy in C_V transfers back to C_{r1} and C_{r2} through S_2 and S_3 during the reimbursement phase. All the switches are off during the harvesting and recovery phases and the variable capacitor is isolated. During the investment phase of the second half cycle, C_{r2} charges the variable capacitor through S_2 and S_4. To avoid discharging C_{r1}, S_3 is off during this phase. The rest of events in this half cycle are the same as harvesting, reimbursement, and recovery phases in the first half cycle. Compared to the harvester in Sect. 3.2.2 that only C_{r1} charges the variable capacitor, C_{r1} and C_{r2} charge the variable capacitor alternately in this harvester.

Fig. 3.18 The sustainable switched-capacitor harvester with two reservoir capacitors, proposed in [3]

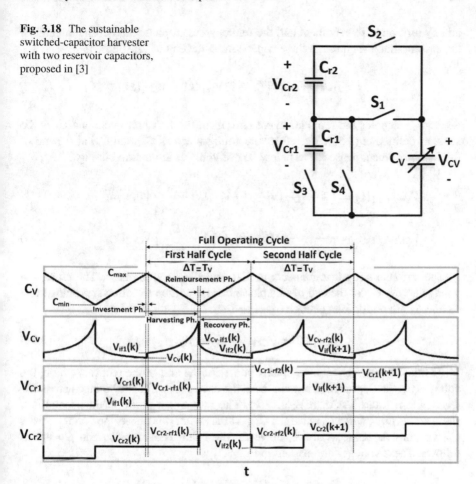

Fig. 3.19 The voltages across C_{r1}, C_{r2}, and C_V of the harvester in Fig. 3.18

3.3.1.1 Closed-Form Expressions

The voltages across C_{r1}, C_{r2}, and C_V are $V_{Cr1}(k)$, $V_{Cr2}(k)$, and $V_{Cv}(k)$, respectively, at the start of the investment phase of the first half cycle. At the end of this phase, $V_{Cr2}(k)$ does not change, while the voltage across C_{r1} and C_V is as follows:

$$V_{if1}(k) = \frac{C_{r1} V_{Cr1}(k) + C_{max} V_{Cv}(k)}{C_{r1} + C_{max}}, \tag{3.91}$$

where $V_{if1}(k)$ denotes the voltage across C_{r1} and C_V at the end of the investment phase of the first half cycle. This voltage is marked on Fig. 3.19.

At the end of the harvesting phase (and the start of the reimbursement phase), the voltages across C_{r1}, C_{r2}, and C_V are $V_{if}(k)$, $V_{Cr2}(k)$, and $nV_{if}(k)$, respectively. S_2

and S_3 turn on at this moment and the reimbursement phase starts. The integral of the current that goes through these capacitors at the end of this phase is

$$\int i_{rr1} dt(k) = C_t \Big((n-1) V_{if1}(k) - V_{Cr2}(k) \Big), \tag{3.92}$$

where i_{rr1} denotes the reimbursement current in the first half cycle and C_t is the same as defined in (3.50). Therefore, the voltages across C_{r1} and C_{r2} at the end of the reimbursement phase are as below. These voltages are marked in Fig. 3.19:

$$V_{Cr1-rf1}(k) = \left(1 + \frac{C_t}{C_{r1}}(n-1)\right) V_{if1}(k) - \frac{C_t}{C_{r1}} V_{Cr2}(k),$$

$$V_{Cr2-rf1}(k) = \frac{C_t}{C_{r2}}(n-1) V_{if1}(k) + \left(1 - \frac{C_t}{C_{r2}}\right) V_{Cr2}(k). \tag{3.93}$$

The variable capacitor connects to C_{r1} and C_{r2} in this phase. Therefore, the voltage across C_V at the end of this phase can be written as follows, provided that S_2 and S_3 are on for long enough (refer to Sect. 3.1.3.1 for details):

$$V_{Cv-rf1}(k) = V_{Cr1-rf1}(k) + V_{Cr2-rf1}(k). \tag{3.94}$$

At this moment, the recovery phase starts. At the end of the recovery phase, the voltages across C_{r1} and C_{r2} remain the same as above, while the voltage across C_V reaches to n times less than $V_{Cv-rf}(k)$. The investment phase in the second half starts at this time, when C_{r2} charges C_V. At the end of this phase, the voltage across C_{r1} remains the same as above; however, the voltages across C_{r2} and C_V are as follows. These voltages are marked in Fig. 3.19:

$$V_{if2}(k) = \frac{C_{r2} V_{Cr2-rf1}(k) + C_{max} \frac{1}{n} V_{Cv-rf1}(k)}{C_{r2} + C_{max}}. \tag{3.95}$$

Using the relation in (3.94) in the above expression:

$$V_{if2}(k) = \frac{1}{\beta + n} V_{Cr1-rf1}(k) + \frac{\beta + 1}{\beta + n} V_{Cr2-rf1}(k), \tag{3.96}$$

where:

$$\beta = \frac{C_{r2}}{C_{min}}. \tag{3.97}$$

The harvesting phase of the second half cycle starts at this moment. At the end of this phase, the voltages across C_{r1} and C_{r2} do not change and are equal to $V_{Cr1-rf}(k)$ and $V_{Cr2-if2}(k)$, respectively. The voltage across C_V reaches to n times more than the above value. The reimbursement phase of the second half cycle starts, and at the end of this phase, the voltages across C_{r1} and C_{r2} are as below. These voltages are marked in Fig. 3.19. Similar procedure as in the above to obtain (3.93) is followed to find these voltages:

$$V_{Cr1-rf2}(k) = \left(1 - \frac{C_t}{C_{r1}}\right) V_{Cr1-rf1}(k) + \frac{C_t}{C_{r1}}(n-1)V_{if2}(k),$$

$$V_{Cr2-rf2}(k) = -\frac{C_t}{C_{r2}} V_{Cr1-rf1}(k) + \left(1 + \frac{C_t}{C_{r2}}(n-1)\right) V_{if2}(k). \qquad (3.98)$$

The voltage across C_V is equal to the sum of the voltages across C_{r1} and C_{r2}; therefore,

$$V_{Cv-rf2}(k) = V_{Cr1-rf2}(k) + V_{Cr2-rf2}(k). \qquad (3.99)$$

The recovery phase starts, and the voltage across C_V reaches to n times less than above at the end of this phase. The voltages across C_{r1} and C_{r2} do not change during this phase. Therefore, at the end of this phase and the start of the investment phase at the $(k+1)$th cycle:

$$V_{Cr1}(k+1) = V_{Cr1-rf2}(k)$$
$$V_{Cr2}(k+1) = V_{Cr2-rf2}(k)$$
$$V_{Cv}(k+1) = \frac{1}{n}V_{Cv-rf2}(k). \qquad (3.100)$$

Replacing the relation in (3.99) in the above expression:

$$V_{Cv}(k+1) = \frac{1}{n}\left(V_{Cr1-rf2}(k) + V_{Cr2-rf2}(k)\right)$$

$$\Rightarrow V_{Cv}(k+1) = \frac{1}{n}\left(V_{Cr1}(k+1) + V_{Cr2}(k+1)\right), \qquad (3.101)$$

where the following expression may be obtained by replacing $k+1$ with k:

$$V_{Cv}(k) = \frac{1}{n}\left(V_{Cr1}(k) + V_{Cr2}(k)\right). \qquad (3.102)$$

Using the above relation in (3.91), $V_{if1}(k)$ can be written as below:

$$V_{if1}(k) = \frac{\alpha+1}{\alpha+n} V_{Cr1}(k) + \frac{1}{\alpha+n} V_{Cr2}(k), \qquad (3.103)$$

where:

$$\alpha = \frac{C_{r1}}{C_{min}}. \tag{3.104}$$

The following steps are carried out to find closed-form expressions for $V_{Cr1}(k)$ and $V_{Cr2}(k)$, assuming that $C_{r1} = C_{r2}$. Following the same steps, the expressions for the case that $C_{r1} \neq C_{r2}$ may be found. However, the reason for this assumption here is that these capacitors play the same role in this harvester. Therefore, the capacitances of these capacitors are selected to be equal in practice:

$$C_{r1} = C_{r2} \Rightarrow \quad \frac{C_{r1}}{C_{min}} = \frac{C_{r2}}{C_{min}} = \alpha = \beta, \quad \frac{C_t}{C_{r1}} = \frac{C_t}{C_{r2}} = \frac{1}{\alpha + 2}. \tag{3.105}$$

Following the sequence of events as above to obtain (3.100), $V_{Cr1}(k + 1)$ and $V_{Cr2}(k + 1)$ are found as below:

$$V_{Cr1}(k + 1) = AV_{Cr1}(k) + BV_{Cr2}(k)$$
$$V_{Cr2}(k + 1) = CV_{Cr1}(k) + DV_{Cr2}(k), \tag{3.106}$$

where:

$$A = \frac{(\alpha + 1)\Big((\alpha + n + 1)(\alpha + 1)(\alpha + n) + n(n - 1)(\alpha + 2)\Big)}{(\alpha + 2)^2(\alpha + n)^2}$$

$$B = \frac{(\alpha + 1)(\alpha + n) + (n - 1)\Big(\alpha^3 + \alpha^2(n + 2) + 3\alpha n + 2n\Big)}{(\alpha + 2)^2(\alpha + n)^2}$$

$$C = \frac{(\alpha + n + 1)(\alpha + 1)\Big((n - 1)\alpha + n\Big)}{(\alpha + 2)^2(\alpha + n)^2}$$

$$D = \frac{(\alpha + n + 1)\Big(\alpha^3 + \alpha^2(n + 2) + 3\alpha n + 2n\Big) - (\alpha + n)}{(\alpha + 2)^2(\alpha + n)^2}. \tag{3.107}$$

To find closed-form formulas for $V_{Cr1}(k)$, a recursive formula should be obtained that only consists of V_{Cr1} terms. The following relation is obtained from the first equation in (3.106):

$$V_{Cr2}(k) = \frac{1}{B}V_{Cr1}(k + 1) - \frac{A}{B}V_{Cr1}(k). \tag{3.108}$$

Replacing k by $k + 1$ in the above expression:

$$V_{Cr2}(k + 1) = \frac{1}{B}V_{Cr1}(k + 2) - \frac{A}{B}V_{Cr1}(k + 1). \tag{3.109}$$

Replacing the above values for $V_{Cr2}(k)$ and $V_{Cr2}(k+1)$ in the second expression of (3.106):

$$V_{Cr1}(k+2) - (A+D)V_{Cr1}(k+1) + (AD - BC)V_{Cr1}(k) = 0. \qquad (3.110)$$

The above recursive formula is linear, homogeneous, and it only consists of the V_{Cr1} terms and it does not contain any V_{Cr2} term. Solving these recursive formulas is explained in Chap. 2. Roots of the above formula are

$$r_1 = \frac{(A+D) + \sqrt{(A-D)^2 + 4BC}}{2}, \qquad r_2 = \frac{(A+D) - \sqrt{(A-D)^2 + 4BC}}{2},$$

$$\qquad (3.111)$$

where r_1 and r_2 are the real distinctive numbers, since B and C are the positive numbers in (3.107) and

$$(A - D)^2 + 4BC > 0. \qquad (3.112)$$

Knowing r_1 and r_2 are the real distinctive numbers, $V_{Cr1}(k)$ is written as follows:

$$V_{Cr1}(k) = \begin{cases} V_{Cr1}(0) & k = 0 \\ a_1 r_1^{k-1} + a_2 r_2^{k-1} & k \geq 1 \end{cases}, \qquad (3.113)$$

where a_1 and a_2 are the constant coefficients and should be found based on initial voltages across the capacitors of this harvester. The expression in (3.110) is obtained considering the relation in (3.102) is true. The initial voltages across C_{r1}, C_{r2}, and C_V are arbitrary, and therefore, the relation in (3.102) may not be true for $k = 0$. Therefore, this relation is true for $k \geq 1$ and $V_{Cr1}(k)$ is written as above for $k \geq 1$.

The initial voltages across C_{r1}, C_{r2}, and C_V are denoted as $V_{Cr1}(0)$, $V_{Cr2}(0)$, and $V_{Cv}(0)$, respectively. Following the steps to find $V_{Cr1}(k+1)$ in (3.100), these voltages are found as follows for $k = 1$:

$$V_{Cr1}(1) = \frac{\alpha}{\alpha + 1} A V_{Cr1}(0) + \frac{\alpha^2(n-2) + \alpha(n-3) - n}{(\alpha + n)(\alpha + 2)^2} V_{Cr2}(0) + \frac{n}{\alpha + 1} A V_{Cv}(0)$$

$$V_{Cr2}(1) = \frac{\alpha}{\alpha + 1} C V_{Cr1}(0) + \frac{\alpha(\alpha + n + 1)(\alpha + 2) + \alpha + n}{(\alpha + n)(\alpha + 2)^2} V_{Cr2}(0) + \frac{n}{\alpha + 1} C V_{Cv}(0).$$

$$\qquad (3.114)$$

Considering $k = 1$ in (3.113):

$$V_{Cr1}(1) = a_1 + a_2. \qquad (3.115)$$

From (3.106), $V_{Cr1}(2)$ is obtained as follows:

$$V_{Cr1}(2) = A V_{Cr1}(1) + B V_{Cr2}(1). \tag{3.116}$$

Considering $k = 2$ in (3.113):

$$V_{Cr1}(2) = a_1 r_1 + a_2 r_2. \tag{3.117}$$

According to the above expressions:

$$\begin{cases} a_1 + a_2 = V_{Cr1}(1) \\ a_1 r_1 + a_2 r_2 = A V_{Cr1}(1) + B V_{Cr2}(1) \end{cases}. \tag{3.118}$$

Solving the above system of equations:

$$a_1 = \frac{(A - r_2) V_{Cr1}(1) + B V_{Cr2}(1)}{r_1 - r_2}, \quad a_2 = \frac{(r_1 - A) V_{Cr1}(1) - B V_{Cr2}(1)}{r_1 - r_2}, \tag{3.119}$$

where A and B are defined in (3.107) and $V_{Cr1}(1)$ and $V_{Cr2}(1)$ are obtained in (3.114). Replacing a_1 and a_2 from the above in (3.113), the closed-form expression of the voltage across C_{r1} is found in (3.113).

Taking similar steps as the above, the closed-form expression for $V_{Cr2}(k)$ is found as below:

$$V_{Cr2}(k) = \begin{cases} V_{Cr2}(0) & k = 0 \\ b_1 r_1^{k-1} + b_2 r_2^{k-1} \end{cases}, \tag{3.120}$$

where $V_{Cr2}(0)$ is the initial voltage across C_{r2}; r_1 and r_2 are found in (3.111) and

$$b_1 = \frac{C V_{Cr1}(1) + (D - r_2) V_{Cr2}(1)}{r_1 - r_2}, \quad b_2 = \frac{-C V_{Cr1}(1) + (r_1 - D) V_{Cr2}(1)}{r_1 - r_2}. \tag{3.121}$$

Finding the closed-form expressions for the voltages across C_{r1} and C_{r2}, the expression for the voltage across C_V is obtained from (3.102) for $k \geq 1$. Therefore,

$$V_{Cv}(k) = \begin{cases} V_{Cv}(0) & k = 0 \\ \dfrac{1}{n} \left(V_{Cr1}(k) + V_{Cr2}(k) \right) & k \geq 1 \end{cases}. \tag{3.122}$$

3.3.1.2   Simulation

This harvester is simulated with the component values of $C_{r1} = C_{r2} = 68\,\text{nF}$, $C_{min} = 4\,\text{nF}$, $C_{max} = 18\,\text{nF}$, and the initial voltages of $V_{Cr1}(0) = 3\,\text{V}$, $V_{Cr2}(0) = 7\,\text{V}$, $V_{Cv}(0) = 1.5\,\text{V}$ across the capacitors. The initial voltages are intentionally

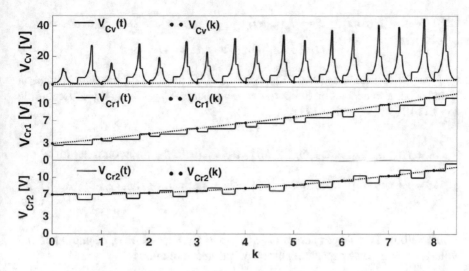

Fig. 3.20 The simulation and values obtained from the closed-form expressions for the voltages across the capacitors of the harvester in Fig. 3.18

chosen to be different to verify the constant coefficients for $V_{Cr1}(k)$ and $V_{Cr2}(k)$ in (3.113) and (3.120). Therefore, the parameters for this simulation are

$$\alpha = 17, \quad n = 4.5, \quad V_{Cr1}(0) = 3 \text{ V}, \quad V_{Cr2}(0) = 7 \text{ V}, \quad V_{Cv}(0) = 1.5 \text{ V}.$$

$$(3.123)$$

Figure 3.20 shows the simulation results of continuous waveforms of the voltages across the capacitors of this harvester ($V_{Cr1}(t)$, $V_{Cr2}(t)$, and $V_{Cv}(t)$). The discrete values for these voltages that are obtained from the closed-form expressions ($V_{Cr1}(k)$, $V_{Cr2}(k)$, and $V_{Cv}(k)$) are shown with circles in this figure. According to the calculations in the previous section, $V_{Cr1}(k)$, $V_{Cr2}(k)$, and $V_{Cv}(k)$ denote the voltages across the capacitors of this harvester at the start of the investment phase of the first half cycle. As can be seen, the closed-form expressions are equal to the simulated waveforms at these moments in each full operating cycle. The discrete values of the voltages are connected with dashed lines to make the comparison easier.

3.3.1.3 The Voltage Dependencies

Reviewing the closed-form expressions for $V_{Cr1}(k)$, $V_{Cr2}(k)$, and $V_{Cv}(k)$, the intention of this section is to find the dependencies of these voltages.

With the assumption in (3.105), all the following parameters only depend on n (the ratio of C_{max} to C_{min}) and α (the ratio of C_{r1} to C_{min}), according to (3.107) and (3.111):

$$A = f_1(n, \alpha), \quad B = f_2(n, \alpha), \quad C = f_3(n, \alpha), \quad D = f_4(n, \alpha),$$
$$r_1 = f_5(n, \alpha), \quad r_2 = f_6(n, \alpha). \tag{3.124}$$

Nonetheless, the constant coefficients in (3.113) and (3.120) not only depend on n and α but also depend on the initial voltages across C_{r1}, C_{r2}, and C_V. According to (3.114), (3.119), and (3.121):

$$a_1 = f_7\left(n, \alpha, V_{Cr1}(0), V_{Cr2}(0), V_{Cv}(0)\right) \quad a_2 = f_8\left(n, \alpha, V_{Cr1}(0), V_{Cr2}(0), V_{Cv}(0)\right)$$
$$b_1 = f_9\left(n, \alpha, V_{Cr1}(0), V_{Cr2}(0), V_{Cv}(0)\right), \quad b_2 = f_{10}\left(n, \alpha, V_{Cr1}(0), V_{Cr2}(0), V_{Cv}(0)\right). \tag{3.125}$$

Therefore, the voltages across C_{r1}, C_{r2}, and C_V depend on n, α, and the initial voltages across these capacitors, based on the above relations:

$$V_{Cr1}(k) = f_{11}\left(n, \alpha, V_{Cr1}(0), V_{Cr2}(0), V_{Cv}(0)\right)$$

$$V_{Cr2}(k) = f_{12}\left(n, \alpha, V_{Cr1}(0), V_{Cr2}(0), V_{Cv}(0)\right)$$

$$V_{Cv}(k) = f_{13}\left(n, \alpha, V_{Cr1}(0), V_{Cr2}(0), V_{Cv}(0)\right). \tag{3.126}$$

The above dependencies for the voltages across the capacitors of this harvester imply that the waveforms for these voltages would be the same for any other component values that result in the same parameters as mentioned in (3.123). Therefore, the voltages across the capacitors of this harvester are exactly the same as depicted in Fig. 3.20 for any of the following component values, provided that the initial voltages across these capacitors are the same as in (3.123):

$set 1 : \; C_{r1} = C_{r2} = 68 \text{ nF}, \quad C_{max} = 18 \text{ nF}, \quad C_{min} = 4 \text{ nF}$

$set 2 : \; C_{r1} = C_{r2} = 68 \text{ pF}, \quad C_{max} = 18 \text{ pF}, \quad C_{min} = 4 \text{ pF}$

$set 3 : \; C_{r1} = C_{r2} = 440 \text{ nF}, \quad C_{max} = 116.5 \text{ nF}, \quad C_{min} = 25.9 \text{ nF}$

$set 4 : \; C_{r1} = C_{r2} = 440 \text{ pF}, \quad C_{max} = 116.5 \text{ pF}, \quad C_{min} = 25.9 \text{ pF}. \tag{3.127}$

The above sets of component values are mentioned as few examples. Any other sets of component values that result in the same parameters as mentioned in (3.123) would be the same as above.

The net generated energy in C_{r1} and C_{r2} at the kth operating cycle of this harvester can be written as follows, considering $C_{r1} = C_{r2}$:

$$E_{net} = \frac{1}{2} C_{r1}\left(V_{Cr1}(k+1)^2 - V_{Cr1}(k)^2 + V_{Cr2}(k+1)^2 - V_{Cr2}(k)^2\right). \tag{3.128}$$

The voltage increase across C_{r1} and C_{r2} from one cycle to the next one is the same for all the component values in (3.127). Therefore, for the example sets in (3.127):

$$\frac{E_{net-set1}}{E_{net-set2}} = \frac{C_{r1-set1}}{C_{r1-set2}} = 1000.$$

Similarly for the rest of the component sets in the above example:

$$E_{net-set3} = 6.4E_{net-set1} = 6470E_{net-set2} = 1000E_{net-set4}. \qquad (3.129)$$

3.3.1.4   Investigating Charge-Depletion Issue

All the voltages across the capacitors of this harvester increase from one cycle to the next one for the component values and the initial voltages of the example that is depicted in Fig. 3.20. Therefore, the large inherent parallel resistors of C_{r1} and C_{r2} do not lead to charge depletion for the component values and initial voltages in this example. In this section, the conditions on the component values and the initial voltages that result in avoiding the charge depletion are investigated.

The parameters that are powered to the $k-1$ term in the expressions of $V_{Cr1}(k)$, $V_{Cr2}(k)$, and $V_{Cv}(k)$ in (3.113), (3.120), and (3.122) are r_1 and r_2. Based on (3.124), these parameters both depend only on n and α. Therefore, evaluating the values of r_1 and r_2 for different values of n and α could be done by plotting r_1 and r_2 versus the changes in n and α. The expressions for $V_{Cr1}(k)$, $V_{Cr2}(k)$, and $V_{Cv}(k)$ consist of two terms: the term that is multiplied by r_1 to the power of $k-1$ and the term that is multiplied by r_2 to the power of $k-1$. Therefore, one of the following three scenarios occurs depending on the values of r_1 and r_2:

Increasing: The voltages across the capacitors increase from one cycle to the next one, if r_1, or/and r_2, is greater than 1 for any values of n and α. In this case, these voltages would be infinite after infinite operating cycles, theoretically: as the voltage across the variable capacitor increases, more energy is extracted from the external energy source. However in practice, the limit in the energy source and the limit in tolerable voltages of the capacitors and the transistor do not allow the voltage increase indefinitely. The expressions for the voltages across the capacitors are derived considering no load or the inherent parallel resistors. In this case, the harvester is capable of avoiding charge depletion if the net generated energy is more than the losses in the parallel resistor and the delivered energy to the load.

Converging: The voltages across the capacitors converge to a voltage, if at least one of r_1 and r_2 is equal to 1 and the other one is less than 1. The convergence to a voltage occurs without any load or the inherent parallel resistors. However in practice, the inherent parallel resistors of the capacitors deplete the charge in these capacitors. The charge depletion occurs regardless of how large these resistors are and this is evident from the analyses for the elementary harvesters in Sect. 3.2.

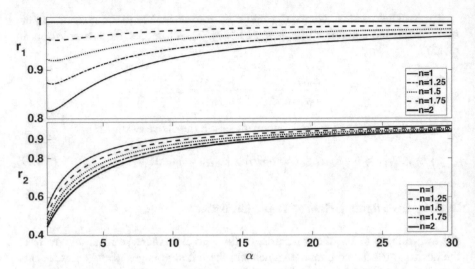

Fig. 3.21 The parameters r_1 and r_2 in (3.113) and (3.120) versus the changes in α and n, when $1 \leq n \leq 2$

Decreasing: The voltages across the capacitors decrease from one cycle to the next one, if both r_1 and r_2 are less than 1 for any values of n and α. In this case, these voltages eventually reach to zero and charge depletion occurs. This happens even without considering the inherent parallel resistors. Therefore in practice, this charge depletion occurs with a faster rate.

Figure 3.21 shows r_1 and r_2 versus α for different values of $1 \leq n \leq 2$. As can be seen, both r_1 and r_2 are less than 1 for any value of $\alpha \geq 1$ and $1 \leq n < 2$. Therefore, this case is relevant to the **decreasing** scenario, as explained in the above, and the voltages across the capacitors eventually reach to zero.

For $n = 2$, the value of r_1 is 1 for any value of α, however $r_2 < 1$. This is relevant to the **converging** scenario, as explained in the above. The voltages across the capacitors of this harvester converge to the following voltages according to the expressions in (3.113), (3.120), and (3.122). However in practice, the inherent parallel resistors deplete the charge in all capacitors:

$$\lim_{k \to \infty} V_{Cr1}(k) = a_1, \quad \lim_{k \to \infty} V_{Cr2}(k) = b_1, \quad \lim_{k \to \infty} V_{Cv}(k) = \frac{1}{n}(a_1 + b_1).$$
$$(3.130)$$

Figure 3.22 shows r_1 and r_2 versus α for different values of n, when $n > 2$. As can be seen, r_1 is always greater than 1, while r_2 is always less than 1. This is relevant with the case **increasing** scenario, as explained in the above. Therefore, the voltages across the capacitors increase from one cycle to the next one for any α, when $n > 2$.

Figure 3.23 shows the simulation and theoretical results of this harvester for the same component values in Sect. 3.3.1.2. However, this figure shows the simulation

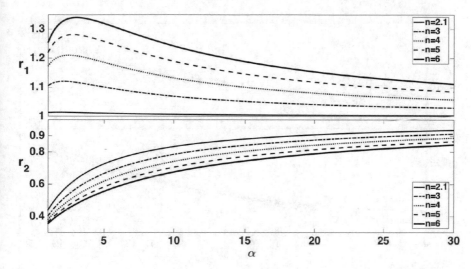

Fig. 3.22 The parameters r_1 and r_2 in (3.113) and (3.120) versus the changes in α and n, when $n > 2$

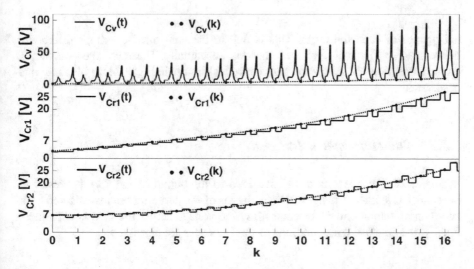

Fig. 3.23 The simulation and values obtained from the closed-form expressions for the voltages across the capacitors of the harvester in Fig. 3.18

for more number of operating cycles to depict the exponential increasing voltages across the capacitors.

As can be seen in Fig. 3.20, the voltage across C_V is asymmetrical in the first and second half cycles of each full cycle. This is due to the different initial voltages across C_{r1} and C_{r2}, in this simulation. Showing the simulation for a more number of cycles in Fig. 3.23, a symmetrical increase in $V_{Cv}(k)$ is seen in the first and second

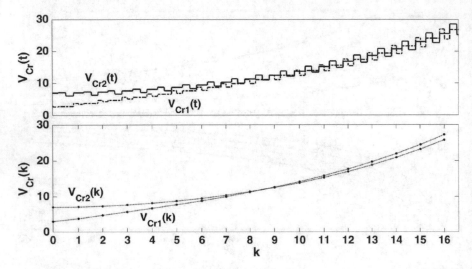

Fig. 3.24 The simulation results and the values from the closed-form expressions for voltages across C_{r1} and C_{r2} of the harvester in Fig. 3.18

half cycles of each full cycle. This is due to the fact that the voltage difference between $V_{Cr1}(k)$ and $V_{Cr2}(k)$ becomes less, eventually. These voltages are shown on the same plots for simulation and theoretical results in Fig. 3.24, to illustrate this occurrence.

3.3.2   The Harvester with Load

In this section, connecting a resistive load to the output of the core harvester in Sect. 3.3.1 is studied. Changes in the voltages of the harvester are investigated first, and then an output circuit for controlling the voltage across the load is explained. Finally, the practical considerations of this harvester are discussed.

3.3.2.1   Maximal Resistive Load

Figure 3.25 shows the harvester in Sect. 3.3.1 with a resistive load. The core harvester is not connected to the load at first. S_L turns on and connects the core of the harvester to R_L, when the voltages across C_{r1} and C_{r2} reach to a desirable value. The number of operating cycles in which S_L turns on is noted as k_L in the rest of this section.

The voltages across the capacitors of this harvester fall into the **increasing** scenario as explained in Sect. 3.3.1.4, when $n > 2$. This means that in each operating cycle the capacitors are charged to a higher voltage. If a load is connected to C_{r1} and

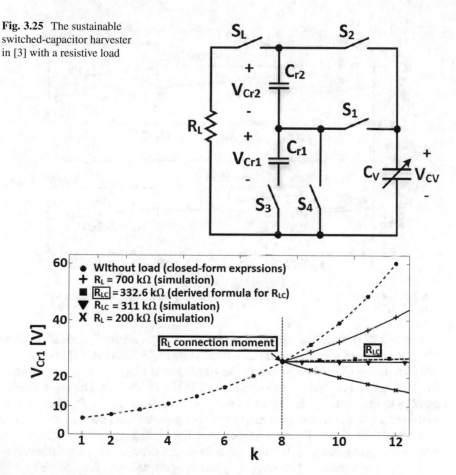

Fig. 3.25 The sustainable switched-capacitor harvester in [3] with a resistive load

Fig. 3.26 Simulation of connecting different resistive loads to the harvester in Fig. 3.25

C_{r2}, it will absorb some charge. Therefore, the voltages across the capacitors may rise slower, they may stay constant, or they may decrease depending upon the value of R_L. These different conditions can be seen in Fig. 3.26 for the voltage across C_{r1}.

The curves with solid lines in Fig. 3.26 are obtained from circuit simulations. The dashed line curve with circle legend is obtained from (3.113) under a no load condition. As shown in this figure, at some arbitrary point in time ($k = k_L = 8$ for this example simulation), the load gets connected to the core harvester. The resistive load that keeps the voltage across C_{r1} and C_{r2} constant is labelled as R_{LC}. The calculation of this parameter is explained in the following paragraphs. The dashed line curve with square legend is drawn based on the calculated value of this parameter, as derived in (3.137). The curves in Fig. 3.26 are obtained for a no load condition as well as four different values of R_L. For these simulations, $C_{r1} = C_{r2} = 100\,\text{nF}$, $C_{max} = 40\,\text{nF}$, $C_{min} = 4\,\text{nF}$, and $T_V = 2\,\text{ms}$.

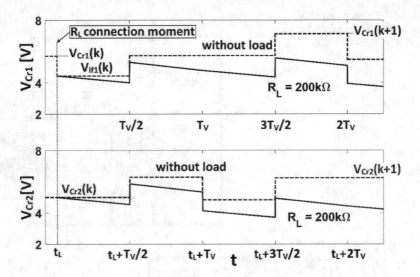

Fig. 3.27 Impact of R_L on the voltage across the capacitors of the harvester in Fig. 3.25

Figure 3.27 shows circuit simulation of node voltages across C_{r1} and C_{r2} of this harvester with and without a load for one operating cycle. This cycle starts at $t = t_L$ and ends at $t = t_L + 2T_V$. Similar to the simulation in Fig. 3.26, it is assumed that S_L is off before this cycle. At the start of this cycle, S_L turns on and connects C_{r1} and C_{r2} to the load, R_L. In Figs. 3.26 and 3.27, the moment that R_L is connected to the output of the harvester is marked. Under no load condition in each quarter of a period ($\Delta T = T_V/2$), the voltages across C_{r1} and C_{r2} are constant. However, when a load is connected to the output, these voltages decrease exponentially during each quarter of a period. Therefore, V_{Cr1} and V_{Cr2} are obtained from the following equations after $T_V/2$, when S_L is closed.

$$V_{Cr1}\left(t_L + \frac{T_V}{2}\right) = V_{Cr1}(t_L) - \frac{1}{C_{r1}} \int_{t_L}^{t_L+T_V/2} i_L(t)dt,$$

$$V_{Cr2}\left(t_L + \frac{T_V}{2}\right) = V_{Cr2}(t_L) - \frac{1}{C_{r2}} \int_{t_L}^{t_L+T_V/2} i_L(t)dt, \qquad (3.131)$$

where i_L is the current that goes through C_{r1}, C_{r2}, and R_L. The integral of this current is as follows based on (3.9):

$$\int_{t_L}^{t_L+T_V/2} i_L(t)dt = C_{12}\Big(V_{Cr1}(t_L) + V_{Cr2}(t_L)\Big), \qquad (3.132)$$

where C_{12} is the series equivalent of C_{r1} and C_{r2} and is obtained in (3.50).

The value of R_L for which these voltages stay constant (R_{LC}) is an important parameter in the behaviour of this harvester. This parameter will now be derived, assuming that S_L is closed when the kth cycle starts and stays closed for the following cycles. The expressions in (3.131) are rewritten as follows, considering the voltages across C_{r1} and C_{r2} at $t = t_L^+$ (after the investment phase of the first half cycle) are $V_{if1}(k)$ and $V_{Cr2}(k)$, respectively:

$$V_{Cr1}\left(t_L + \frac{T_V}{2}\right) = V_{if1}(k_L) - \left(V_{if1}(k_L) + V_{Cr2}(k_L)\right)\frac{C_{r2}}{C_{r1}+C_{r2}}\left(1 - e^{-\frac{T_V}{2R_L C_{12}}}\right),$$

$$V_{Cr2}\left(t_L + \frac{T_V}{2}\right) = V_{Cr2}(k_L) - \left(V_{if1}(k_L) + V_{Cr2}(k_L)\right)\frac{C_{r1}}{C_{r1}+C_{r2}}\left(1 - e^{-\frac{T_V}{2R_L C_{12}}}\right).$$

$$(3.133)$$

Same expressions as the above can be written to find $V_{Cr1}(t_L+T_V)$ and $V_{Cr2}(t_L+T_V)$, where $V_{if1}(k)$ and $V_{Cr2}(k)$ in the above are replaced with the voltages across C_{r1} and C_{r2} after the reimbursement phase of the first half cycle. Taking same steps, the voltages across C_{r1} and C_{r2} at $t = t_L + 3T_V/2$ may be obtained. Following the sequence of events and the fact that the voltages across C_{r1} and C_{r2} decrease during each $T_V/2$ time period, the voltages across C_{r1} and C_{r2} may be obtained after one full cycle in the presence of R_L. Alternatively, these voltages may be estimated as follows at the end of this full cycle:

$$V_{Cr1-RL}(k_L + 1) \approx V_{Cr1}(k_L + 1) -$$

$$\left(V_{Cr1}(k_L)+V_{Cr2}(k_L)+V_{Cr1}(k_L+1)+V_{Cr2}(k_L+1)\right)\frac{C_{r2}}{C_{r1}+C_{r2}}\left(1 - e^{-\frac{T_V}{R_L C_{12}}}\right),$$

$$V_{Cr2-RL}(k_L + 1) \approx V_{Cr2}(k_L + 1) -$$

$$\left(V_{Cr1}(k_L)+V_{Cr2}(k_L)+V_{Cr1}(k_L+1)+V_{Cr2}(k_L+1)\right)\frac{C_{r1}}{C_{r1}+C_{r2}}\left(1 - e^{-\frac{T_V}{R_L C_{12}}}\right),$$

$$(3.134)$$

where $k = k_L$ denotes the cycle in which S_L turns on. V_{Cr1-RL} and V_{Cr2-RL} are the voltages across C_{r1} and C_{r2} at the $(k_L + 1)$th cycle in the presence of R_L. In the above estimations, the terms $V_{Cr1}(k_L + 1)$ and $V_{Cr2}(k_L + 1)$ are the voltages across C_{r1} and C_{r2} at the $(k_L + 1)$th cycle when no load is present. The second terms in these estimations represent a similar decrease as in (3.133). In these terms, the voltages across C_{r1} and C_{r2} are assumed to be $V_{Cr1}(k_L)$ and $V_{Cr2}(k_L)$ for the first half cycle and $V_{Cr1}(k_L + 1)$ and $V_{Cr2}(k_L + 1)$ for the second half cycle. This way, the previously derived expressions for $V_{Cr1}(k)$ and $V_{Cr2}(k)$ in (3.113) and (3.120) are used to find R_{LC}. To find the R_{LC} at which the voltages across C_{r1} and C_{r2} are constant, $V_{Cr1-RL}(k_L + 1)$ and $V_{Cr2-RL}(k_L + 1)$ should be set equal to $V_{Cr1}(k_L)$ and $V_{Cr2}(k_L)$:

$$V_{Cr1-RL}(k_L + 1) = V_{Cr1}(k_L), \quad V_{Cr2-RL}(k_L + 1) = V_{Cr2}(k_L). \tag{3.135}$$

It is assumed that S_L remains closed after this cycle as depicted in Fig. 3.26. Therefore, the calculated value of R_{LC} from the consideration in (3.135) keeps the voltages across C_{r1} and C_{r2} constant after the R_L connection moment ($k = k_L$).

According to Fig. 3.22, if $n > 2$, the first term with an exponent of $k-1$ in (3.113) and (3.120) is more than one, while the second term is less than one: $r_1 > 1, r2 < 1$. Therefore, for a large enough k ($k_L \gg 1$), the second terms in the expressions of $V_{Cr1}(k)$ and $V_{Cr2}(k)$ can be ignored; hence, these expressions are estimated as below:

$$k_L \gg 1 \Rightarrow \begin{cases} V_{Cr1}(k_L) \approx a_1 r_1^{k_L-1} \\ \\ V_{Cr2}(k_L) \approx b_1 r_1^{k_L-1}. \end{cases} \tag{3.136}$$

Replacing the above values for $V_{Cr1}(k_L)$, $V_{Cr2}(k_L)$, $V_{Cr1}(k_L+1)$, and $V_{Cr2}(k_L+1)$ in the first expression in (3.134) and using the condition in (3.135), R_{LC} is obtained as follows:

$$R_{LC} = \frac{-2T_V}{C_{r1} \ln \left(1 - \frac{2a_1(r_1-1)}{(a_1+b_1)(r_1+1)} \right)}, \tag{3.137}$$

where C_{r1} is equal to C_{r2} and, therefore, $C_{r2}/(C_{r1} + C_{r2})$ and $C_{r1}/(C_{r1} + C_{r2})$ are both 0.5. Moreover, the series equivalent of C_{r1} and C_{r2} is $C_{r1}/2$. Based on the above expression, R_{LC} does not depend on k, when the assumptions in (3.136) are reasonable. Therefore, the value of R_{LC} is obtained from the above expression, regardless of the moment that S_L turns on. In this case, a load resistor with the value equal to R_{LC} is able to keep the output voltage constant at any arbitrary voltage. Therefore, the output voltage may be controlled based on the moment that S_L turns on when $R_L = R_{LC}$. In practice, R_L is determined based on the application and may be different from R_{LC}. The solution to this is explained in Sect. 3.3.2.4. R_{LC} is used as a useful parameter in calculating the net generated energy and understanding the behaviour of this harvester in the rest of this section. Therefore, noting the connection of R_{LC} to the output of this harvester does not mean that the harvester is not able to provide power to any other resistor. This is discussed in more detail in Sect. 3.3.2.4.

Figure 3.28 shows the simulation of the harvester in Fig. 3.25, where R_{LC} is calculated from (3.137). In this figure, the moment that S_L turns on and connects C_{r1} and C_{r2} to the load is marked. The voltages across C_{r1} and C_{r2} are increasing before this moment, and after this moment, these voltages are kept constant due to the connection of the output to R_{LC}. This parameter is calculated from (3.137) for the component values of the simulation in Fig. 3.26: $C_{r1} = C_{r2} = 100\,\text{nF}$, $C_{max} = 40\,\text{nF}$, $C_{min} = 4\,\text{nF}$, $V_{Cr1}(0) = V_{Cr2}(0) = 5\,\text{V}$, $V_{Cv}(0) = 3\,\text{V}$, and $T_V = 2\,\text{ms}$. The R_{LC} connection moment is at $t = 30\,\text{ms}$ in this simulation: this moment is at the start of the second half cycle of the eighth cycle. The calculated R_{LC} is $332.6\,\text{k}\Omega$

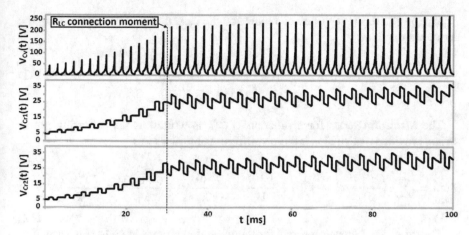

Fig. 3.28 Simulation of the harvester in Fig. 3.25, when S_L connects at the kth cycle

and is slightly different from the R_{LC} that is obtained from simulation ($311\,\text{k}\Omega$), according to Fig. 3.26. This difference is due to the estimations in deriving (3.137). As can be seen in Fig. 3.28, the voltages across C_{r1} and C_{r2} continue to increase with a negligible rate due to this difference.

In practice, C_{r1} and C_{r2} may be selected to be much larger than C_{min}. This way, the amount of storable energy in C_{r1} and C_{r2} at a specific voltage is increased, and the ripple in the voltage across the load is decreased. In this case, the following assumption is valid:

$$C_{r1} \gg C_{min} \Rightarrow \alpha \to \infty. \tag{3.138}$$

The values for A, B, C, and D are as below, considering the above assumption in (3.107):

$$A \approx D \approx 1 - \frac{1}{\alpha} \qquad B \approx C \approx \frac{n-1}{\alpha}. \tag{3.139}$$

Reasonably, the initial voltages across C_{r1} and C_{r2} may be considered equal. As can be seen in Figs. 3.23 and 3.24, the voltages across C_{r1} and C_{r2} eventually equate, even if they are not equal at the beginning. From this moment on (the moment that the voltages across C_{r1} and C_{r2} equate), the operation of the harvester is the same as if the harvester started with same initial voltages across C_{r1} and C_{r2}. With these assumptions:

$$\begin{cases} \alpha \to \infty \\ V_{Cr1}(0) = V_{Cr2}(0) \end{cases} \Rightarrow a_1 = b_1. \tag{3.140}$$

R_{LC} in (3.137) is rewritten as below for the case that $\alpha \gg 1$, considering the above assumptions:

$$R_{LC-\alpha\infty} = \frac{-2T_V}{C_{r1} \ln \left(1 - \frac{r_1 - 1}{r_1 + 1}\right)}. \tag{3.141}$$

The Maclaurin series for a function, $f(x)$, is written as the following power series, provided that $f(x)$ is infinitely differentiable at $x = 0$:

$$f(x) = \sum_{j=0}^{j=\infty} \frac{f^{(j)}(0)}{j!} x^j = f(0) + \frac{f'(0)}{1!} x + \frac{f''(0)}{2!} x^2 + \frac{f'''(0)}{3!} x^3 + \frac{f^{(4)}(0)}{4!} x^4 + \dots \tag{3.142}$$

Therefore, the Maclaurin series for the natural logarithm in (3.141) is written as below:

$$\ln\left(1 - \frac{r_1 - 1}{r_1 + 1}\right) = -\frac{r_1 - 1}{r_1 + 1} - \frac{1}{2}\left(\frac{r_1 - 1}{r_1 + 1}\right)^2 - \frac{1}{3}\left(\frac{r_1 - 1}{r_1 + 1}\right)^3 - \frac{1}{4}\left(\frac{r_1 - 1}{r_1 + 1}\right)^4 - \dots, \tag{3.143}$$

where

$$\frac{r_1 - 1}{r_1 + 1} \ll 1 \Rightarrow \ln\left(1 - \frac{r_1 - 1}{r_1 + 1}\right) \approx -\frac{r_1 - 1}{r_1 + 1}. \tag{3.144}$$

Using the values of A, B, C, and D from (3.139) and the above approximation for the natural logarithm in (3.145):

$$R_{LC-\alpha\infty} = \frac{4T_V}{C_{min}(n - 2)}. \tag{3.145}$$

3.3.2.2 Energy Calculations

In this harvester, the voltages across C_{r1} and C_{r2} increase, when the output of the harvester is not connected to any load. A resistive load equal to R_{LC} keeps the output voltage constant at any moment that S_L turns on. Considering that S_L turns on at the kth cycle, the output voltage of the harvester remains constant at $V_{Cr1}(k) + V_{Cr2}(k)$, when $R_L = R_{LC}$. This is shown in Fig. 3.28. The harvested energy not only compensates the energy that is obtained from C_{r1} and C_{r2} during the investment phases, but also provides the energy for keeping the voltage across the load constant. Therefore, the net generated energies in C_{r1} and C_{r2} are zero, and the net generated energy in this harvester is calculated as below:

$$E_{net} = \frac{\left(V_{Cr1}(k_L) + V_{Cr2}(k_L)\right)^2}{R_{LC}} T_V. \tag{3.146}$$

E_{net} is calculated for a single half cycle (T_V) in the above expression. With assumptions in (3.140), the expression in (3.145) may be used for R_{LC} in the above expression; hence,

$$E_{net} = \frac{1}{4} C_{min}(n-2)\left(V_{Cr1}(k_L) + V_{Cr2}(k_L)\right)^2. \tag{3.147}$$

The above estimation for E_{net} when $C_{r1}, C_{r2} \gg C_{min}$, implies that the net generated energy is only positive for $n > 2$ in this harvester.

To find the deliverable energy (E_{del}) and the total conduction losses, the QV diagram of this harvester is plotted in Fig. 3.29. This figure shows the QV diagram for all the first half cycles after the $k = k_L$ cycle, when S_L turns on. Point **a** marks the start of the investment phase. The voltage across C_V is obtained from (3.122) at this moment. The end of the investment phase is marked by point **b**, when $V_{Cv} = V_{if1}(k_L)$, and is expressed in (3.103). Calculating the enclosed area in this diagram, the deliverable energy in the first half cycle of each full cycle is obtained as follows:

$$E_{del} = \frac{1}{2} C_{max}(n-1)\left(V_{if1}(k_L) - V_{Cv}(k_L)\right)\left(V_{if1}(k_L) + V_{Cv}(k_L)\right). \tag{3.148}$$

Replacing the values of $V_{if1}(k = k_L)$ and $V_{Cv}(k = k_L)$ from (3.103) and (3.122) in the above:

Fig. 3.29 The QV diagram of the harvester in Fig. 3.25

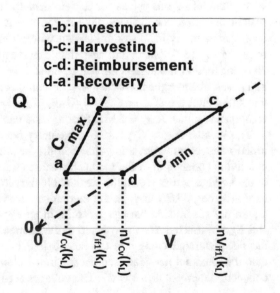

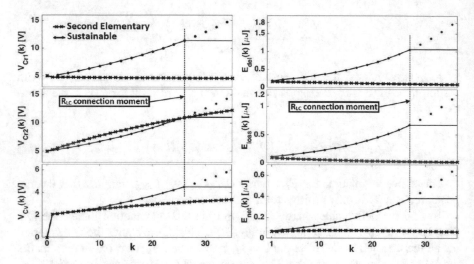

Fig. 3.30 Comparison of the V_{Cr1}, V_{Cr2}, V_{Cv}, E_{del}, E_{loss}, and E_{net} between the second elementary harvester in Fig. 3.12 and the sustainable harvester in Fig. 3.25

$$E_{del} = \frac{1}{2} C_{min} \frac{n-1}{n(\alpha+n)^2} \Big(\alpha(n-1)V_{Cr1}(k_L) - \alpha V_{Cr2}(k_L) \Big)$$

$$\Big((\alpha(n+1) + 2n)V_{Cr1}(k_L) + (\alpha + 2n)V_{Cr2}(k_L) \Big). \qquad (3.149)$$

Figure 3.30 compares the second elementary harvester in Fig. 3.12 and the sustainable harvester in Fig. 3.25. The horizontal axis in the plots of this figure is k, the number of full cycles for the elementary harvester. Each full cycle for the sustainable harvester in these plots is equivalent to two increments in k. Signals for the elementary harvester are shown with no load, while the moment that R_{LC} connects to the output of the sustainable harvester is marked in this figure. The R_{LC} connection moment is chosen to be at $k = 27$; however, this moment could be chosen at any other arbitrary k. The component values for both harvesters in this simulation are $C_{r1} = C_{r2} = 40\,\text{nF}$, $C_{max} = 5\,\text{nF}$, $C_{min} = 1\,\text{nF}$. With these component values, $R_{LC} = 3\,\text{M}\Omega$ for the sustainable harvester when $T_V = 2\,\text{ms}$.

The voltage across C_{r1} in the elementary harvester stays constant at about its initial voltage; however, this voltage in the sustainable harvester keeps increasing before the connection of the load. This causes a faster and an exponential increase in the voltage across C_V in the sustainable harvester, where this voltage converges to $V_{if}(0)$ based on (3.64) in the elementary harvester. The higher voltage across C_V in the sustainable harvester results in an exponential increase in E_{del}, E_{loss}, and E_{net} of this harvester compared to the elementary harvester. In this simulation, the net generated energy in the sustainable harvester is about five times more than the maximal net generated energy in the elementary harvester, when the R_{LC} connection moment is at $k = 27$. The voltages across the capacitors and the energies in these harvesters could be compared if the R_{LC} connection moment would be different.

3.3.2.3 Efficiency Optimization

The energy efficiency is defined as the ratio of the net generated energy in a harvester to the energy that is harvested from the energy source:

$$\eta = \frac{E_{net}}{E_h}, \tag{3.150}$$

where E_{net} is the net generated energy in the harvester, which is a fraction of the harvested energy, E_h, from the energy source.

The harvested energy for the charge-constraint and voltage-constraint switching schemes (E_{hC} and E_{hV}) is explained in Sects. 2.2.1 and 2.2.2, respectively. In this harvester, the voltages across C_V are $V_{if1}(k)$ and $nV_{if1}(k)$ at the start and end of the harvesting phase, respectively. This is depicted in Fig. 3.29. Therefore,

$$E_{hC} = \frac{1}{2}C_{max}(n-1)V_{if1}{}^2(k_L),$$

$$\Rightarrow E_{hC} = \frac{1}{2}C_{max}(n-1)\left(\frac{\alpha+1}{\alpha+n}V_{Cr1}(k_L) + \frac{1}{\alpha+n}V_{Cr2}(k_L)\right)^2, \tag{3.151}$$

where $V_{if1}(k_L)$ is obtained from (3.103), at $k = k_L$.

The net generated energy in this harvester for the case that $C_{r1}, C_{r2} \gg C_{min}$ ($\alpha \to \infty$) is obtained in (3.147). Considering this assumption in the above expression for $V_{if1}(k_L)$, the energy efficiency is found as below:

$$\eta = \frac{E_{net}}{E_{hC}} = \frac{\frac{1}{4}C_{min}(n-2)\left(V_{Cr1}(k_L) + V_{Cr2}(k_L)\right)^2}{\frac{1}{2}C_{max}(n-1)V_{if1}{}^2(k_L)}$$

$$\xrightarrow{\alpha\to\infty} \eta \approx \frac{(n-2)\left(V_{Cr1}(k_L) + V_{Cr2}(k_L)\right)^2}{2n(n-1)V_{Cr1}{}^2(k_L)}. \tag{3.152}$$

S_L turns on when the voltages across C_{r1} and C_{r2} reach to the desired value. Therefore, the voltages across C_{r1} and C_{r2} do not depend on n. Optimizing the energy efficiency, the optimal n is found as below:

$$\frac{d\eta}{dn} = 0 \Rightarrow n_{opt} = 2 + \sqrt{2} \approx 3.4. \tag{3.153}$$

The efficiency in (3.152) may be estimated as below, when $\alpha \to \infty$ ($C_{r1}, C_{r2} \gg C_{min}$):

$$\eta \approx \frac{2(n-2)}{n(n-1)}, \tag{3.154}$$

where the following assumptions are taken into account:

$$V_{if1}(k_L) \approx V_{Cr1}(k_L), \quad V_{Cr1}(k_L) \approx V_{Cr2}(k_L). \tag{3.155}$$

The first approximation in the above is evident from (3.103), where $\alpha \to \infty$. The second approximation is obtained with the assumptions in (3.136) and (3.140) that were used to obtain R_{LC} for $\alpha \to \infty$.

The energy efficiency of this harvester in (3.154) only depends on n. Therefore, it is plotted versus n in Fig. 3.31. Since the net generated energy is only positive if $n > 2$, the plot starts at $n = 2.1$. The energy efficiency is optimized at 0.34 when $n_{opt} = 3.4$.

Considering the assumptions in (3.155), E_{hC} and E_{net} are plotted in Fig. 3.32 versus n, when $C_{max} = 5\,\text{nF}$ and $V_{Cr1}(k_L) = 5$. In these plots, C_{max} is considered a constant value; therefore, different values of n change C_{min}. This is a practical assumption, since an upper limit exists for C_{max} depending on what technology is used to fabricate the variable capacitor. The harvested energy in the first half cycle of each full cycle for different values of n is shown in Fig. 3.32a. This parameter increases linearly when n increases. The net generated energy for the same duration in each cycle versus n is shown in Fig. 3.32b. To plot this figure, the expression in (3.147) is used. There is limit on achievable C_{max} in a specific technology and in a defined volume, in practice. Therefore, C_{min} is replaced with C_{max}/n in this expression to plot E_{net} versus changes in n. As can be seen, E_{net} also increases when n increases. However, the increase rate in this parameter eventually reaches to zero. Therefore, the energy efficiency increases at first when n increases, but it starts decreasing at some point and eventually reaches to zero when $n \to \infty$. This is evident from the changes in E_{hC} and E_{net} when n increases. E_{net} converges to

Fig. 3.31 The energy efficiency of the harvester in Fig. 3.25 when $C_{r1}, C_{r2} \gg C_{min}$ ($\alpha \to \infty$)

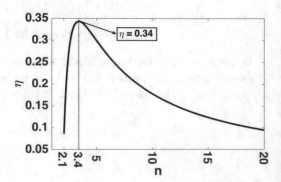

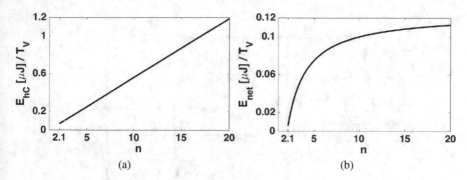

Fig. 3.32 (a) The harvested energy and (b) the net generated energy in a duration of T_V for the harvester in Fig. 3.25

the following value when $n \to \infty$:

$$lim_{n\to\infty} E_{net} = \lim_{n\to\infty} \frac{1}{4}\frac{C_{max}}{n}(n-2)\big(2V_{Cr1}(k_L)\big)^2 = C_{max}V_{Cr1}^2(k_L). \qquad (3.156)$$

3.3.2.4   Voltage Control Mechanism

This harvester is capable of generating any desired voltage, provided the energy source supplies the required energy and the switches tolerate this voltage. However, when a load is connected to the output, care must be exercised such that the generator can sustain its voltage. If the value of R_L is chosen to be equal to R_{LC}, then the output voltage will be constant and only depends on the moment S_L turns on. However, the value of R_L may change due to many different factors. If R_L exceeds R_{LC}, the output voltage will rise eventually, and if R_L drops below R_{LC}, the output voltage will reach to zero, eventually.

This harvester needs a voltage control mechanism to keep the harvester running properly. The harvester with a voltage control mechanism is shown in Fig. 3.33. The value of R_L should be chosen to be smaller than R_{LC}. The control circuit senses the output voltage and connects the load by turning S_L on, when this voltage reaches a desired high value (V_H). As soon as S_L is closed, the output voltage decreases since $R_L < R_{LC}$. When the voltage of the generator drops below a specified value (V_L), the control circuit opens S_L. Therefore, the output voltage increases until the control circuit closes S_L again. This procedure continues, and the output voltage of the harvester stays within a specified range (comprised between V_H and V_L).

Figure 3.34 shows the simulation result of the circuit in Fig. 3.33. In this simulation, $R_L = 100\,\text{k}\Omega$; the other parameters are the same as those used for simulation of Fig. 3.27. With these parameters, $R_{LC} = 332.6\,\text{k}\Omega$. Therefore, a $100\,\text{k}\Omega$ load decreases the output voltages whenever S_L is closed. The high and low threshold voltages (V_L and V_H) are chosen to be 10 and 20 V, respectively. The

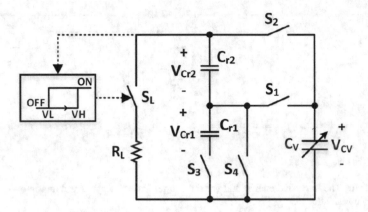

Fig. 3.33 Presented output voltage control mechanism for the harvester in Fig. 3.25 in [3] in the presence of a resistive load

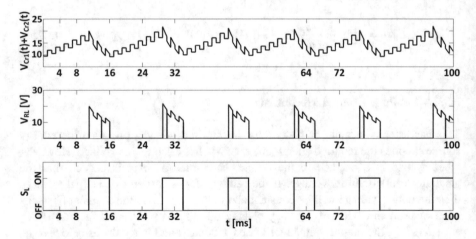

Fig. 3.34 The simulation of the harvester in Fig. 3.33, where $V_H = 20$ V and $V_L = 10$ V and $R_L = 100$ kΩ

voltage control mechanism turns S_L off, when the output voltage reaches to V_L. The output voltage during this period follows $V_{Cr1}(k) + V_{Cr2}(k)$, where these voltages are expressed in (3.113) and (3.120): these expressions are obtained when no load is present.

The energy output of most of the kinetic energy sources, such as moving limbs, varies strongly. As an example, the energy output using the knee joint movement is higher when a person runs. Moreover, the harvester may need to avoid applying any suppressing pressure on the energy source. As an example, in proposing an energy harvesting system for human diaphragm, care must be exercised in the amount harvested energy. Based on the structure of the variable capacitor, the energy source, and the electromechanical coupling, variations in the available energy may result in that the variable capacitor changes between C_{max} and a value higher than C_{min}. Therefore, n, R_{LC}, and the output voltage become lower, higher, and lower than expected, respectively. Depending on the energy source, the variations related to the energy source and its impact on n may need to be compensated.

This harvester is capable of providing a flexible output energy. The simulations in Fig. 3.30 illustrate this flexibility, where the net generated energy in this harvester could be adjusted by determining the output voltage. This is especially of interest when the available energy in an energy source varies in time. In this case, the harvester needs to be capable of adjusting the amount of its harvested energy, either to maximize the net generated energy or not to put any suppressing pressure on the energy source. This harvester is able to deliver the same amount of power to any load less than R_{LC}, if the voltage control mechanism keeps V_L and V_H constant. In this case, the time that S_L is closed decreases for smaller R_L values. R_{LC} depends on the parameters of the generator (as defined in (3.145)), and a reasonably high value can be chosen to extend the allowable range of loads.

3.3.2.5 Start-Up and Fault Prevention Mechanism

Battery-free switching electrostatic harvesters, including the harvester in Fig. 3.33, suffer from the following problem. If by any means, e.g. an unwanted short circuit, the capacitor charges are depleted, the harvester stops working. This issue becomes more important when the harvester is not easily accessible. To eliminate this problem, a start-up circuit is added to this harvester. The proposed circuit in [3] is shown in Fig. 3.35. The start-up circuit comprises a battery (B_1) and a diode (D_1). The voltage of the battery (V_{B1}) is smaller than the normal voltage of C_{r1}. Therefore, the diode is off during the normal operation of the harvester. However, if the charge in C_{r1} is depleted due to an unwanted event, the diode conducts and charges C_{r1} to a voltage equal to $V_{B1} - V_{F-D1}$, where V_{F-D1} is the forward voltage drop across D_1. Under this condition, the control circuit disconnects R_L from the harvester, since the output voltage of the generator is less than V_L. The output voltage of the harvester rises until it reaches V_H. Therefore, the output voltage of the harvester may rise to V_H and when this happens, the control circuit turns S_L on again. During the normal operation of the generator, D_1 disconnects B_1 from C_{r1}.

3.3.3 The Harvester for Battery Charging

Figure 3.36 shows the core harvester in Sect. 3.3.1 that is used to charge a battery. This circuit is proposed in [4]. The voltages across the capacitors of this harvester and the current that goes through D_1 are shown in Fig. 3.37. A full cycle and the phases of operation in one half cycle are shown in this figure. The voltage across C_V increases during the harvesting phase. As can be seen, the reimbursement phase in this harvester has two parts. Part 1 of the reimbursement phase starts when

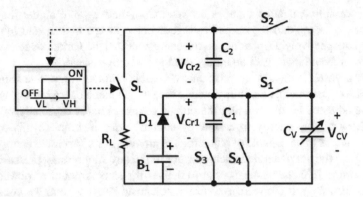

Fig. 3.35 The harvester in Fig. 3.33 with a start-up and fault prevention circuit

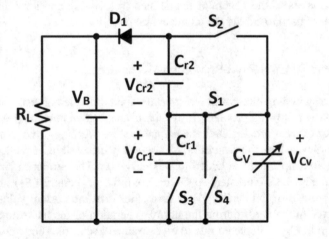

Fig. 3.36 Sustainable core harvester in Fig. 3.18 for charging a battery

the voltage across C_V becomes equal to the sum of the voltages across C_{r1} and C_{r2}. S_2 turns on at the start of this part, and C_V connects to C_{r1} and C_{r2}. Since the capacitance of C_V is still decreasing at this moment, the voltages across C_V, C_{r1}, and C_{r2} all increase together. The current that goes through C_{r1}, C_{r2}, and C_V during this part of the reimbursement phase is plotted as i_{rr-p1} in Fig. 3.37. This part continues until when the voltages across C_V and the sum of the voltages across C_{r1} and C_{r2} become equal to V_B. Part 2 of the reimbursement phase starts at this moment, D_1 turns on, and C_V connects to the battery. The capacitance of C_V continues decreasing until the end of the harvesting phase: D_1 stays on until the end of the harvesting phase. Therefore, a charging current goes through D_1 and the battery during this time. i_{D1} shows this current in Fig. 3.37.

The recovery phase starts at the end of the reimbursement phase. The voltage across C_V decreases from V_B to V_B/n at the end of this phase. In the second half cycle, C_{r2} charges C_V during the investment phase. The sum of the voltages across

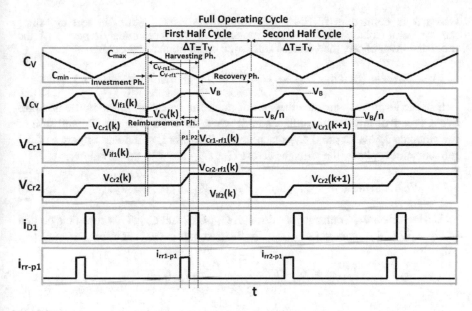

Fig. 3.37 Voltages across the capacitors and the battery charging current of the harvester in Fig. 3.36

C_{r1} and C_{r2} is less than V_B at the end of this phase; therefore, part 1 and part 2 of the reimbursement phase are followed in the same order as in the previous half cycle.

3.3.3.1 Closed-Form Expressions

Considering $V_{Cr1}(k)$, $V_{Cr2}(k)$, and $V_{Cv}(k)$ as the voltages across C_{r1}, C_{r2}, and C_V at the start of the investment phase of the first half cycle of the kth cycle (as depicted in Fig. 3.37):

$$V_{Cv}(k) = \frac{V_B}{n}. \tag{3.157}$$

The above relation is obtained from the operation of this harvester as explained in Sect. 3.3.3. C_{r1} connects to C_V through S_1 and S_3 during the investment phase. Assuming $C_{r1} = C_{r2}$, the voltage across C_{r1} and C_V ($V_{if1}(k)$) is obtained as follows at the end of this phase:

$$V_{if1}(k) = \frac{\alpha}{\alpha + n} V_{Cr1}(k) + \frac{1}{\alpha + n} V_B, \tag{3.158}$$

where α is defined in (3.104). The voltage across C_V starts increasing, since the harvesting phase begins at this moment. At the beginning of part 1 of the reimbursement phase, the voltages across the capacitors are as below:

$$V_{Cr1-rs1}(k) = V_{if1}(k), \quad V_{Cr2-rs1}(k) = V_{Cr2}(k), \quad V_{Cv-rs1}(k) = V_{if1}(k) + V_{Cr2}(k),$$
(3.159)

where the term "rs1" denotes the start of the reimbursement phase in the first half cycle. Part 1 of this phase continues until when the voltage across C_V and sum of the voltages across C_{r1} and C_{r2} reach to V_B. Therefore, at the end of part 1 of the reimbursement phase, the voltages across C_V, C_{r1}, and C_{r2} are as below:

$$V_{Cr1-rf1}(k) + V_{Cr2-rf1}(k) = V_B \qquad V_{Cv-rf1}(k) = V_B.$$
(3.160)

During this time, a current goes through C_V, C_{r1}, and C_{r2}. This current is plotted in Fig. 3.37 and is marked as i_{rr1-p1} for the first half cycle. Therefore,

$$V_{Cr1-rf1}(k) = V_{if1}(k) + \frac{1}{C_{r1}} \int i_{rr1-p1}(t)dt,$$

$$V_{Cr2-rf1}(k) = V_{Cr2}(k) + \frac{1}{C_{r2}} \int i_{rr1-p1}(t)dt.$$
(3.161)

According to (3.160) and (3.161):

$$V_{if1}(k) + V_{Cr2}(k) + \frac{1}{C_{12}} \int i_{rr1-p1}(t)dt = V_B$$

$$\Rightarrow \int i_{rr1-p1}(t)dt = C_{12}\Big(V_B - V_{if1}(k) - V_{Cr2}(k)\Big),$$
(3.162)

where C_{12} is the series equivalent capacitances of C_{r1} and C_{r2} and is defined in (3.50). Knowing the integral of i_{rr1-p1}, the voltages across C_{r1} and C_{r2} at the end of part 1 of the reimbursement phase are found as below:

$$V_{Cr1-rf1}(k) = V_{if1}(k) + \frac{C_{12}}{C_{r1}}\Big(V_B - V_{if1}(k) - V_{Cr2}(k)\Big),$$

$$V_{Cr2-rf1}(k) = V_{Cr2}(k) + \frac{C_{12}}{C_{r2}}\Big(V_B - V_{if1}(k) - V_{Cr2}(k)\Big).$$
(3.163)

Replacing C_{12} with $C_{r1}/2$ in this harvester, since $C_{r1} = C_{r2}$:

$$V_{Cr1-rf1}(k) = 0.5\Big(V_B + V_{if1}(k) - V_{Cr2}(k)\Big),$$

$$V_{Cr2-rf1}(k) = 0.5\Big(V_B + V_{Cr2}(k) - V_{if1}(k)\Big).$$
(3.164)

These voltages are marked in Fig. 3.37. Part 2 of the reimbursement phase starts and the battery keeps the voltage across C_V constant at V_B, while a charging current goes through C_V and the battery. The reimbursement phase continues until the capacitance of C_V reaches to C_{min}. D_1 turns off at this moment and the recovery phase. At the end of the recovery phase, the capacitance of C_V is maximal (C_{max}) and the voltage across it is V_B/n.

At the start of the investment phase in the second half cycle of the kth cycle, the voltages across C_V, C_{r1}, and C_{r2} are V_B/n, $V_{Cr1-rf1}(k)$, $V_{Cr2-rf1}(k)$, respectively. During this phase, S_2 and S_4 turn on, while S_1 and S_3 are off. Therefore, C_{r2} connects to C_V, and the voltage across these capacitors at the end of this phase is

$$V_{if2}(k) = \frac{\alpha}{\alpha + n} V_{Cr2-rf1}(k) + \frac{1}{\alpha + n} V_B, \tag{3.165}$$

where $V_{Cr2-rf1}(k)$ is obtained in (3.164). The harvesting phase starts, and the voltage across C_V starts increasing. Part 1 of the reimbursement phase in the second half cycle starts when the voltage across C_V becomes equal to the sum of the voltages across C_{r1} and C_{r2}. Therefore,

$$V_{Cr1-rs2}(k) = V_{Cr1-rf1}(k), \qquad V_{Cr2-rs2}(k) = V_{if2}(k),$$

$$V_{Cv-rs2}(k) = V_{Cr1-rf1}(k) + V_{if2}(k), \tag{3.166}$$

where "rs2" denotes the start of the reimbursement phase in the second half cycle. Following the same steps to find the integral of i_{rr1-p1} for the first half cycle in (3.162), the integral of the current that goes through C_{r1}, C_{r2}, and C_V during part 1 of the reimbursement phase of the second half cycle is found as follows:

$$\int i_{rr2-p1}dt = C_{12}\Big(V_B - V_{Cr1-rf1}(k) - V_{if2}(k)\Big), \tag{3.167}$$

and therefore, using similar expressions to (3.163) and considering that $C_{r1} = C_{r2} = 2C_{12}$, the voltages across C_{r1} and C_{r2} at the end of part 1 of the reimbursement phase are found. As can be seen in Fig. 3.37, these voltages do not change until the end of the kth cycle. Therefore, these voltages are equal to the voltages across C_{r1} and C_{r2} at the $(k+1)$th cycle:

$$V_{Cr1}(k+1) = 0.5\Big(V_B + V_{Cr1-rf1}(k) - V_{if2}(k)\Big),$$

$$V_{Cr2}(k+1) = 0.5\Big(V_B + V_{if2}(k) - V_{Cr1-rf1}(k)\Big). \tag{3.168}$$

Replacing the value of $V_{if1}(k)$ from (3.158) in (3.164), $V_{Cr1-rf1}(k)$ and $V_{Cr2-rf1}(k)$ are obtained based on $V_{Cr1}(k)$, $V_{Cr2}(k)$, and V_B. Using these values in

(3.165) and (3.168), $V_{Cr1}(k+1)$ in the above expression is written as follows based on $V_{Cr1}(k)$, $V_{Cr2}(k)$, and V_B;

$$V_{Cr1}(k+1) = 0.25\frac{\alpha(2\alpha+n)}{(\alpha+n)^2}V_{Cr1}(k)-$$

$$0.25\frac{2\alpha+n}{\alpha+n}V_{Cr2}(k) + 0.25\frac{(\alpha+n)(2\alpha+3n)-n}{(\alpha+n)^2}V_B. \qquad (3.169)$$

Sum of the voltages across C_{r1} and C_{r2} at the end of the reimbursement phase of the first or the second half cycle is equal to V_B. This can be verified by adding these voltages in (3.164) for the first half cycle. Similarly, adding these voltages in the above expression gives the same result. Therefore,

$$V_{Cr1}(k+1) + V_{Cr2}(k+1) = V_B. \qquad (3.170)$$

Replacing $k+1$ with k in the above expression:

$$V_{Cr1}(k) + V_{Cr2}(k) = V_B \Rightarrow V_{Cr2}(k) = V_B - V_{Cr1}(k). \qquad (3.171)$$

Replacing the above value for $V_{Cr2}(k)$ in (3.169):

$$V_{Cr1}(k+1) = 0.25\frac{(2\alpha+n)^2}{(\alpha+n)^2}V_{Cr1}(k) + 0.25\frac{2n(\alpha+n)-n}{(\alpha+n)^2}V_B. \qquad (3.172)$$

The above expression is a linear non-homogeneous recursive formula. Referring to Sect. 2.4.2, this expression is solved and the result is written as below.

$$V_{Cr1}(k) = V_{Cr1-g}(k) + V_{Cr1-p}(k)$$

$$\Rightarrow V_{Cr1}(k) = c1\left(1 - \frac{n}{2(\alpha+n)}\right)^{2(k-1)} + \frac{2(\alpha+n)-1}{4\alpha+3n}V_B \qquad k \geq 1,$$

$$(3.173)$$

where $c1$ is a constant coefficient and should be found from the initial conditions, i.e. the initial voltages across C_{r1}, C_{r2}, and C_V. The above expression is only valid for $k \geq 1$, since the relation in (3.157) is considered to be true in the kth cycle. However, the initial voltage across C_V may not follow this condition. To find $c1$ in the above, it is required to obtain the voltage across C_{r1} at $k = 1$. Following the sequence of events in a cycle as explained, the following formula is found for the voltage across C_{r1} after the first cycle:

$$V_{Cr1}(1) = 0.25 \left(\frac{\alpha(2\alpha + n)}{(\alpha + n)^2} V_{Cr1}(0) - \frac{2\alpha + n}{\alpha + n} V_{Cr2}(0) + \right.$$

$$\left. \frac{n(2\alpha + n)}{(\alpha + n)^2} V_{Cv}(0) + \frac{2(\alpha + n) + n - 2}{\alpha + n} V_B \right), \quad (3.174)$$

where $V_{Cr1}(0)$, $V_{Cr2}(0)$, and $V_{Cv}(0)$ are the initial voltages across C_{r1}, C_{r2}, and C_V, respectively. Using the value of $V_{Cr1}(1)$ in (3.173) when $k = 1$:

$$(3.173) \overset{k=1}{\Longrightarrow} V_{Cr1}(1) = c1 + \frac{2(\alpha + n) - 1}{4\alpha + 3n} V_B$$

$$\Rightarrow c1 = V_{Cr1}(1) - \frac{2(\alpha + n) - 1}{4\alpha + 3n} V_B. \quad (3.175)$$

Therefore, the voltage across C_{r1} is expressed as below:

$$V_{Cr1}(k) = \begin{cases} V_{Cr1}(0) & k = 0 \\ c1\left(1 - \dfrac{n}{2(\alpha + n)}\right)^{2(k-1)} + \dfrac{2(\alpha + n) - 1}{4\alpha + 3n} V_B & k \geq 1 \end{cases}, \quad (3.176)$$

where $c1$ is defined in (3.175). Using the expression in (3.171), the voltage across C_{r2} is expressed as below:

$$V_{Cr2}(k) = \begin{cases} V_{Cr2}(0) & k = 0 \\ -c1\left(1 - \dfrac{n}{2(\alpha + n)}\right)^{2(k-1)} + \dfrac{2\alpha + n + 1}{4\alpha + 3n} V_B & k \geq 1 \end{cases}. \quad (3.177)$$

Based on (3.157), the voltage across C_V is

$$V_{Cv}(k) = \begin{cases} V_{Cv}(0) & k = 0 \\ \dfrac{V_B}{n} & k \geq 1 \end{cases}. \quad (3.178)$$

The obtained expressions for the voltages across C_{r1}, C_{r2}, and C_V are compared with the simulation of this harvester in Fig. 3.38. The results in this figure are shown for $C_{r1} = C_{r2} = 40\,\text{nF}$, $C_{max} = 10\,\text{nF}$, $C_{min} = 1\,\text{nF}$, $V_{Cr1}(0) = 2.7\,\text{V}$, $V_{Cr2}(0) = 7.3\,\text{V}$, $V_{Cv}(0) = 0.8\,\text{V}$, and $V_B = 10\,\text{V}$. The solid lines in this figure show the simulation results, while the values from the expressions in (3.176) (3.177), and (3.178) are marked with scattered dots. These dots are connected with dashed lines to illustrate how these voltages evolve based on the obtained expressions. As can be seen in this figure, the initial voltage across C_V is 0.8 V; however, this voltage is equal to V_B/n for the rest of the cycles: $V_{Cv}(k) = 1\,\text{V}$ for $k \geq 1$. The initial voltages across C_{r1} and C_{r2} are chosen not to be equal to verify the obtained

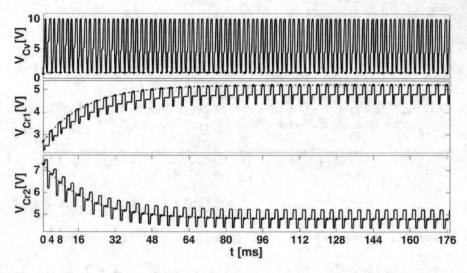

Fig. 3.38 The comparison between a simulation of the harvester in Fig. 3.36 and the obtained expressions for the voltages across C_{r1}, C_{r2}, and C_V

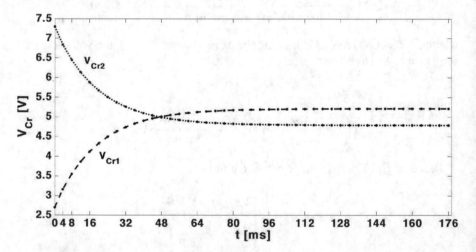

Fig. 3.39 The evolution of the voltages across C_{r1} and C_{r2} in Fig. 3.38, based on the obtained expression

expressions. As can be seen, the values from the expressions are in complete agreement with the simulation results. To show how the voltages across C_{r1} and C_{r2} evolve, these voltages are shown on the same plot in Fig. 3.39.

The voltages across C_{r1} and C_{r2} in this harvester eventually reach to their final (steady-state) values. This can be seen in Figs. 3.38 and 3.39. The final values of these voltages are obtained as follows from (3.176) and (3.177):

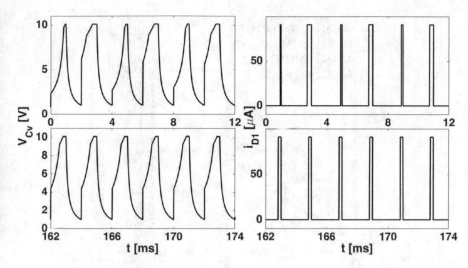

Fig. 3.40 The voltage across C_V and the charging current through the battery in Fig. 3.38 for the first and last few operating cycles

$$V_{Cr1-f} = \lim_{k\to\infty} V_{Cr1}(k) = \frac{2(\alpha + n) - 1}{4\alpha + 3n}V_B, \qquad V_{Cr2-f} \lim_{k\to\infty} V_{Cr2}(k) = \frac{2\alpha + n + 1}{4\alpha + 3n}V_B,$$

$$(3.179)$$

where V_{Cr1-f} and $V_{Cr2\,f}$ refer to the final (steady-state) voltages across C_{r1} and C_{r2}. These values are obtained noting the term that is powered to $2(k-1)$ in (3.176) and (3.177) is less than one. Therefore, this term reaches to zero when $k \to \infty$.

3.3.3.2 Energy Calculations

The voltage across C_V and the charging current that goes through the battery are plotted in Fig. 3.40 for the first few cycles (during the transient period) and for the last few cycles of the run simulation (during the steady-state period). In transient, these signals are asymmetrical in the first and second half cycles, due to the different initial voltages across C_{r1} and C_{r2}. However, these signals become symmetrical in the first and second half cycles when the voltages across C_{r1} and C_{r2} reach to the steady-state period. These values are obtained in (3.179). The net generated energy, the deliverable energy, and the conduction losses in this harvester are calculated in the following paragraphs based on the final (steady-state) voltages across C_{r1} and C_{r2}.

Figure 3.41 shows the QV diagram of this harvester, when the voltages across C_{r1} and C_{r2} reach to their final (steady-state) values. Point **a** marks the start of the investment phase, when $V_{Cv}(k) = V_B/n$ and $C_V = C_{max}$. At this moment, C_{r1} connects to C_V: the voltage across C_V reaches to V_{if1-f} at the end of the investment phase. Point **b** marks the end of the investment phase. The expression for $V_{if1}(k)$ is obtained in (3.158). Using the final value of the voltage across C_{r1} (V_{Cr1-f} in

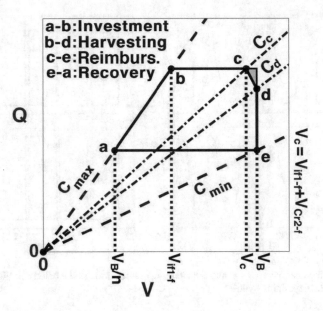

Fig. 3.41 The QV diagram of the harvester in Fig. 3.36 when the voltages across C_{r1} and C_{r2} reach to their final (steady-state) values

(3.179)), in this expression:

$$V_{if1-f} = \frac{\alpha}{\alpha + n} V_{Cr1-f} + \frac{1}{\alpha + n} V_B = \frac{2\alpha + 3}{4\alpha + 3n} V_B. \tag{3.180}$$

During the harvesting phase, the voltage across C_V increases and reaches to V_c, while C_V is isolated from the rest of the circuit. This point is marked as **c** in Fig. 3.41, and V_c is defined on this figure. At this moment, the capacitance of C_V is C_c and is obtained as follows:

$$C_c = \frac{V_{if1-f}}{V_{if1-f} + V_{Cr2-f}} C_{max}, \tag{3.181}$$

where the charge conservation law is used for C_V, since this capacitor is isolated from the rest of the circuit. At this moment, S_2 turns on and C_V connects to C_{r1} and C_{r2}. This is the start of the reimbursement phase (part 1) overlapping with the harvesting phase. The capacitance of C_V continues decreasing until the voltage across it reaches to V_B. The capacitance of C_V is equal to C_d when the voltage across it reaches to V_B. This point is marked as **d** in Fig. 3.41. Moving from **c** to **d**, the equivalent capacitor of C_{r1}, C_{r2}, and C_V is isolated from the rest of the circuit, since D_1 is off. Writing the charge conservation law for this equivalent capacitor, C_d is obtained as follows:

$$V_B = \frac{C_c + \frac{C_{r1}}{2}}{C_d + \frac{C_{r1}}{2}}\left(V_{if1-f} + V_{Cr2-f}\right)$$

$$\Rightarrow C_d = \frac{(n+2)\alpha + 3n}{4\alpha + 3n}C_{min}, \tag{3.182}$$

where it is considered that $C_{r1} = C_{r2}$. At this moment, the voltage across C_V is equal to V_B; therefore, D_1 turns on and C_V connects to the battery. This is the start of part 2 of the reimbursement phase, overlapping with the harvesting phase. The battery keeps the voltage across C_V constant at V_B, while the capacitance of C_V continues decreasing until it reaches to C_{min}. The end of the harvesting and reimbursement phases is marked as **e**. The capacitance of C_V starts changing from its minimal value towards its maximal value at this point. At the end of the recovery phase (point **a**), C_V is ready for the second half cycle.

Following the sequence of events and calculating C_c and C_d, the enclosed area in Fig. 3.41 is calculable. This area is equal to the deliverable energy (E_{del}). Subtracting the greyed area in this figure from the area of the big trapezoid:

$$E_{del} = \frac{\left(V_B - \frac{V_B}{n} + V_B - V_{if1-f}\right)\left(C_{max}V_{if1-f} - C_{max}\frac{V_B}{n}\right)}{2} -$$
$$\frac{(V_B - V_{if1-f} - V_{Cr2-f})(C_{max}V_{if1-f} - C_d V_B)}{2}. \tag{3.183}$$

Replacing the values of V_{if1-f}, V_{Cr2-f}, and C_d in the above expression:

$$E_{del} = \frac{\alpha(n-2)\Big((2\alpha+n)(3n-2)+2n(n-1)\Big)}{n(4\alpha+3n)^2}C_{min}V_B^2. \tag{3.184}$$

The voltages across C_{r1} and C_{r2} eventually reach to their final values, as calculated in (3.179). Therefore, these voltages do not change from one cycle to the next one and the net generated energy in these capacitors is zero. However, the net generated energy in the battery is positive. This energy is calculated based on the integral of the current that goes through the battery during part 2 of the reimbursement phase, when D_1 is on. This current is shown in Fig. 3.40, for the final voltages across C_{r1} and C_{r2}. The integral of this current during the first half cycle is

$$\int i_{D1} = (C_d - C_{min})V_B. \tag{3.185}$$

Details of finding the above expression are explained in Sect. 4.1.3. Based on the above:

$$E_{net} = (C_d - C_{min})V_B{}^2 = \frac{(n-2)\alpha}{4\alpha + 3n}C_{min}V_B{}^2. \tag{3.186}$$

Apparently, the net generated energy in this harvester is positive only if $n > 2$, based on the above expression. The following expression is obtained for the net generated energy in this harvester, when $\alpha \gg n$:

$$C_{r1}, C_{r2} \gg C_{min} \Rightarrow \alpha \gg n \Rightarrow E_{net} = \frac{1}{4}C_{min}(n-2)V_B{}^2. \tag{3.187}$$

Knowing the values of E_{del} and E_{net} in (3.184) and (3.186), the conduction losses in this harvester are calculated as below:

$$E_{loss-tot} = E_{del} - E_{net}$$

$$E_{loss-tot} = \frac{2\alpha(n-2)^2(\alpha+n)}{n(4\alpha+3n)^2}C_{min}V_B{}^2. \tag{3.188}$$

3.3.3.3 Efficiency

The energy efficiency of an electrostatic harvester is defined in (3.150). The net generated energy of the harvester in Fig. 3.36 is calculated in (3.186). To calculate the energy efficiency of this harvester, the harvested energy (E_h) in each half cycle is calculated in this section. The below steps should be followed, noting that the harvesting phase of this harvester overlaps with its reimbursement phase and the reimbursement phase consists of two parts.

The voltage across C_V increases from V_{if1-f} to V_c during the harvesting phase, while C_V is isolated from the rest of the circuit, Therefore,

$$E_{h-bc} = \frac{1}{2}C_c V_c{}^2 - \frac{1}{2}C_{max}V_{if1-f}{}^2, \tag{3.189}$$

where E_{h-bc} is the harvested energy when C_V status moves from point **b** to point **c** in Fig. 3.41. Using the values of V_{if1-f} and V_{Cr2-f} in (3.181), C_c and V_c (the voltage of C_V at point **c** in Fig. 3.41) are obtained as follows:

$$C_c = \frac{2\alpha+3}{4\alpha+n+4}C_{max}, \qquad V_c = \frac{4\alpha+n+4}{4\alpha+3n}V_B. \tag{3.190}$$

Replacing the above values for C_c, V_c in (3.189):

$$E_{h-bc} = \frac{1}{2}C_{min}V_B{}^2\frac{n(2\alpha+3)(2\alpha+n+1)}{(4\alpha+3n)^2}. \tag{3.191}$$

In Sect. 4.1.4, how energy is harvested and transferred from a variable capacitor while it is connected to a storage component is detailed. For a better understanding of how the harvested energy during the reimbursement phase of this harvester is calculated in the following paragraphs, it is recommended that this section is reviewed first.

At the point **c**, part 1 of the reimbursement phase starts during which C_V is connected to C_{r1} and C_{r2}. This part of the reimbursement phase ends at point **d**. During this, the amount of stored energy in the variable capacitor is reduced by the following amount:

$$E_{Cv-cd} = \frac{1}{2}C_c V_c^2 - \frac{1}{2}C_d V_B^2, \tag{3.192}$$

where E_{Cv-cd} is the amount of reduced stored energy in C_V when the status of C_V moves from point **c** to point **d** in the QV diagram. Using the values of C_c, V_c, and C_d in (3.190) and (3.182) in the above expression:

$$E_{Cv-cd} = \frac{1}{2}C_{min} V_B^2 \frac{(n-2)(4\alpha^2 - n\alpha - 6n)}{(4\alpha + 3n)^2}. \tag{3.193}$$

The voltage across the series equivalent capacitor of C_{r1} and C_{r2} increases from V_c to V_B during this part of the reimbursement phase. Therefore,

$$E_{Cr-cd} = \frac{1}{2}\frac{C_{r1}}{2}V_B^2 - \frac{1}{2}\frac{C_{r1}}{2}V_B^2$$
$$\Rightarrow E_{Cr-cd} = 2\alpha C_{min} V_B^2 \frac{(n-2)(2\alpha + n + 1)}{(4\alpha + 3n)^2}, \tag{3.194}$$

where E_{Cr-cd} is the amount of energy that is received by the series equivalent capacitor of C_{r1} and C_{r2} during part 1 of the reimbursement phase, where $C_{r1} = C_{r2}$. Considering a similar block diagram to Fig. 4.4, the amount of harvested energy during part 1 of the reimbursement phase is obtained as below, when conduction losses are negligible:

$$E_{h-cd} = E_{Cr-cd} - E_{Cv-cd}$$
$$\Rightarrow E_{h-cd} = \frac{1}{2}C_{min} V_B^2 \frac{(n-2)\left(4\alpha^2 + (5n+4)\alpha + 6n\right)}{(4\alpha + 3n)^2}, \tag{3.195}$$

where E_{h-cd} is the harvested energy when the status of C_V moves from point **c** to point **d** in Fig. 3.41.

Part 2 of the reimbursement phase starts at point **d** in Fig. 3.41 during which C_V connects to the battery. Following similar steps as above, the harvested energy during this part of the reimbursement phase is as follows. Details of these steps are explained in Sect. 4.1.4.

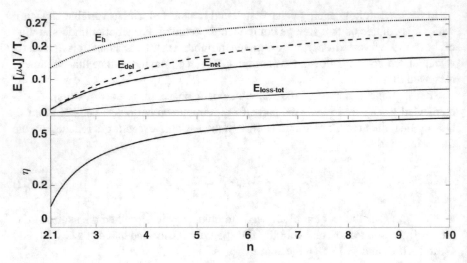

Fig. 3.42 E_h, E_{del}, E_{net}, $E_{loss-tot}$, and the energy efficiency in the harvester of Fig. 3.36 versus n, when $C_{r1} = C_{r2} = 40\,\text{nF}$, and $C_{max} = 10\,\text{nF}$

$$E_{h-de} = \frac{1}{2}(C_d - C_{min})V_B{}^2$$

$$\Rightarrow E_{h-de} = \frac{1}{2}C_{min}V_B{}^2\frac{(n-2)\alpha}{4\alpha + 3n}. \tag{3.196}$$

According to the above explanations, the total harvested energy during the harvesting phase in this harvester is

$$E_h = E_{h-bc} + E_{h-cd} + E_{h-de}$$

$$\Rightarrow E_h = \frac{1}{2}C_{min}V_B{}^2\frac{4(3n-4)\alpha^2 + 2(5n^2 - 2n - 4)\alpha + 9n(n-1)}{(4\alpha + 3n)^2}. \tag{3.197}$$

Knowing the value of E_h in the above and E_{net} in (3.186), the energy efficiency of this harvester is obtained:

$$\eta = \frac{E_{net}}{E_h} = \frac{2\alpha(n-2)(4\alpha + 3n)}{4(3n-4)\alpha^2 + 2(5n^2 - 2n - 4)\alpha + 9n(n-1)}. \tag{3.198}$$

Figure 3.42 shows the harvested energy of this harvester versus changes in n for the case that $C_{r1} = C_{r2} = 40\,\text{nF}$ and $C_{max} = 10\,\text{nF}$. In this plot, C_{max} is constant when n changes to simulate the limit on C_{max} (in a specific technology and a defined volume) in practice. Therefore, C_{min} changes when n changes. The deliverable energy, the net generated energy, and the total conduction losses are shown on the plot. This plot shows these energies in half cycle (equivalent to T_V). The energy efficiency is also plotted in this figure. As can be seen, the energy

Fig. 3.43 The discrete
NMOS reverse current
control test circuit

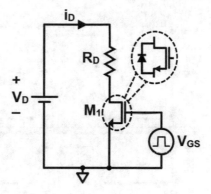

efficiency increases for higher values of n. Therefore, it is desirable to increase n in this harvester. However, the efficiency and the net generated energy do not increase much after $n = 8$ for the component parameters in this simulation. The maximal energy efficiency of this harvester is obtained as follows:

$$C_{r1},\ C_{r2} \gg C_{min} \xRightarrow{\alpha \gg n} \eta = \frac{2(n-2)}{3n-4}. \tag{3.199}$$

Based on the above expression, the energy efficiency of this harvester is maximized when n is maximal; therefore,

$$\lim_{n\to\infty} \eta = \lim_{n\to\infty} \frac{2(n-2)}{3n-4} \approx 0.66. \tag{3.200}$$

3.4 Circuit Implementation

Discrete or integrated transistors may be used in implementing the switches in the switched-capacitor harvesters. The practical considerations for each of these categories are discussed in the following subsections.

3.4.1 Discrete Transistors

Discrete transistors intrinsically have a reverse diode between the drain and the source terminals. Therefore, current passes through a discrete transistor when a reverse voltage is applied across it, regardless of its Gate–Source voltage. Figure 3.43 shows a simulation setup in LTspice with a pulse generator connected to the gate of the generic NMOS model. A resistor is placed in series with the transistor, and a DC voltage source, V_D, is connected to this resistor and the Source

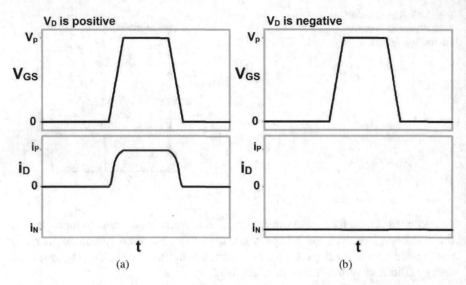

Fig. 3.44 The current that goes through an NMOS transistor with a pulse voltage applied to its Gate–Source when (**a**) a positive voltage is across its Drain–Source and (**b**) a negative voltage is across its Drain–Source

of the NMOS. Figure 3.44 shows the simulation of this setup for a positive and a negative V_D. In this figure, i_D is the current that goes through the Drain–Source of the transistor with the direction specified in Fig. 3.43. The Gate–Source pulse controls i_D when V_D is positive; therefore, $i_D = 0$ when $V_{GS} = 0$. However, a negative current goes through the Drain–Source of the transistor when V_D is negative. The negative current means that this current goes in the opposite direction that is marked for i_D in Fig. 3.43. The applied control signal to the Gate–Source of the transistor does not stop this current, and the transistor conducts in an undesired manner. This is due to the intrinsic diode between the Drain and the Source of the discrete transistor. This diode is shown in a more detailed symbol of the discrete transistor in Fig. 3.43.

Figure 3.45 shows a discrete implementation of the harvester in Fig. 3.36. To have a common source node for all the transistors, S_2 and S_3 of the circuit in Fig. 3.36 are swapped with C_{r2} and C_{r1} without affecting the functionality of the circuit. Having a common source for all of the transistors simplifies the implementation of the control circuit, which should provide voltages between the Gates and Sources of the transistors. Considering the intrinsic diode in discrete transistors, diodes are used in series with the transistors to avoid undesired current going through the transistor. The currents in S_2 and S_3 (in Fig. 3.36) are bidirectional, while the currents in S_1 and S_4 are always in one direction. Therefore, S_2 and S_3 are implemented with complementary NMOS and PMOS switches, while S_1 and S_4 are implemented with a single MOSFET.

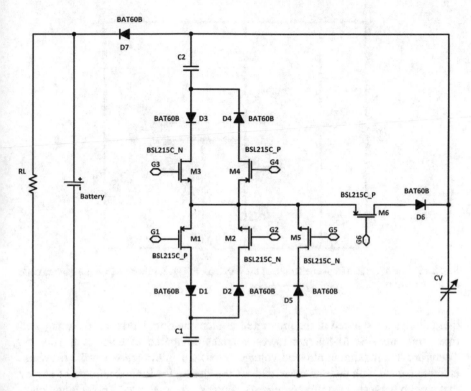

Fig. 3.45 The discrete implementation of the harvester in Fig. 3.36 with a common source node for all transistors

3.4.2   Integrated Transistors

Integrating the switches and the control circuit in one chip is desirable in applications with limited available volume. This way, an Application Specific Integrated Circuit (ASIC) may be designed to improve the efficiency of the control circuit for the core harvester. Figure 3.46 shows an implementation of the harvester in Fig. 3.36 with integrated transistors. All of the transistors have a common source in this implementation simplifying the control circuit. The bidirectional switches S_2 and S_3 in Fig. 3.36 are implemented with a single integrated MOSFET, provided that a triple-well CMOS technology is employed. In this technology, the current passing through the transistor in an undesired direction may be avoided by connecting the bulk of NMOS transistors to the lowest voltage of the circuit. Equivalently, the bulk of PMOS transistors should be connected to the highest voltage of the circuit. This way, the intrinsic diode between Drain and Source would be reversed biased at all times, and the current through this diode would be blocked.

The series diodes in the discrete implementation of Fig. 3.45 are eliminated in the integrated implementation of Fig. 3.46. Therefore, power losses related to

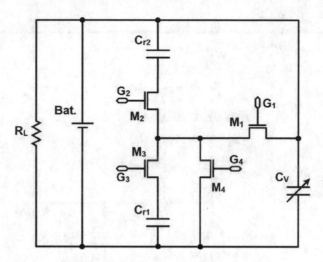

Fig. 3.46 The integrated implementation of the harvester in Fig. 3.36 with a common source node for all transistors

these diodes are omitted in the integrated implementation. However, the integrated transistors are able to tolerate lower voltages compared to discrete transistors. Therefore, the maximum allowed voltage across C_V is limited when the harvester is implemented with integrated switches. Considering the limited amount of energy that can be harvested from many energy sources, the integrated implementation is still advantageous in energy harvesting from these sources.

3.4.3 Control Circuit

In this chapter, analysing the core of sustainable electrostatic harvesters is presented, and the behaviour of these harvesters when a resistive or battery load connects to their output is detailed. These harvesters require a control block that generates the gate signals of the transistors. The harvesters in this chapter are in the category of **S:S** harvesters: both switching events at the start of the investment and the reimbursement phases are synchronous (this term is detailed in Sect. 2.1.5.2). Therefore, the control block should detect both of the moments when the capacitance of the variable capacitor is maximal and minimal. A block diagram including all the necessary components in this control block is presented in Sect. 5.4.1.2.

An Application Specific Integrated Circuit (ASIC) is designed for the control block in [5] for an **A:S** electrostatic harvester. In this control circuit, a level shifter generates the gate pulse of a transistor based on the output of the decision logic that detects the moments that the capacitance of the variable capacitor is minimal. The reported consumption power for this control logic inside the ASIC is 24 nW, which is achieved by biasing the transistors in sub-threshold region. A reservoir capacitor supplies the ASIC and receives a part of the harvested power in each operating cycle in order not to be depleted. The required control circuit for the harvesters in this section has to detect both of the moments that the capacitance of the variable capacitor is maximal and minimal. The power consumption of the required control circuit for these harvesters would not exceed a few hundreds of nW. Therefore, the whole system is practical if the generated energy exceeds this required energy for the control circuit.

3.5 Applications and Simulations

Structure of a promising variable capacitor in energy harvesting from low frequency long amplitude movements is detailed in this section. Next, energy harvesting from knee joint and diaphragm muscle is detailed, considering the features of this variable capacitor. The simulation of these two harvesting systems and an estimation of their sizes are presented later.

3.5.1 A Variable Capacitor

The capacitance of a simple variable capacitor with a stationary plate and a moving plate linked to a linear kinetic energy source changes from maximal to minimal and back to maximal again once for a full movement cycle. However, the variable capacitor that is proposed in [6] is capable of increasing the frequency of changes in the capacitance for linear kinetic energy sources. Figure 3.47 shows the structure of this variable capacitor that may be fabricated using the microfluidics technology. Figure 3.47a shows an implementation of this variable capacitor with four pairs of electrodes. The moving body consists of conducting and dielectric liquid regions alternately and is mechanically linked to a linear movement kinetic energy source. This moving part moves in a channel between the pairs of electrodes. If the conducting liquid regions are aligned with electrode pairs, the capacitance is maximal. The capacitance is minimal, if the dielectric liquid regions are aligned with the electrode pairs. Figure 3.47b shows a zoomed in view of a pair of electrodes, when a conducting liquid is aligning with them. The conducting liquid results in forming two relatively large capacitors in series, since a very thin dielectric layer is separating the formed electrodes. However, the electrodes of the variable capacitor are separated by the width of the channel when dielectric liquids are aligned with them. Therefore, the capacitance of this implementation changes from maximal to minimal and back to maximal again eight times for a full linear movement cycle. The frequency of changes may be increased by increasing the number of conductor pairs, conducting liquid regions, and dielectric liquid regions.

The dielectric layers between the channel and the electrodes could be as thin as 20 nm according to [6]. Therefore, a variable capacitor with a maximal capacitance

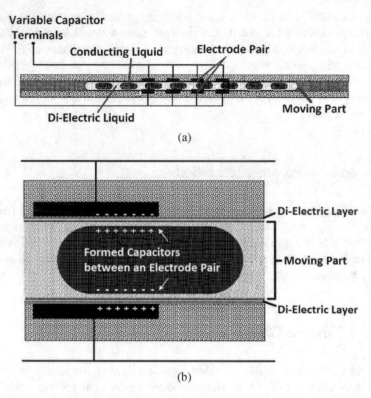

Fig. 3.47 The variable capacitor proposed in [6]. (**a**) An implementation of this variable capacitor with four pairs of conductors (**b**) Zoomed in on one pair of conductors

of 39 μF is claimed to be implementable in a relatively tiny volume. Using this variable capacitor, it is also possible to increase the frequency of a linear kinetic energy source.

3.5.2   Energy Harvesting from Knee Joint

Energy harvesting from limbs is a notable candidate in powering wearable devices to either eliminate battery or prolong the battery life span. Figure 3.48 shows an implementation of an electrostatic harvester for energy harvesting from knee joint movement in [3] using the variable capacitor in Fig. 3.47. In [3], the circuit in Sect. 3.3.2 is used for the harvester in Fig. 3.48.

The normal movement frequency of knee joint should be considered as low as 1 Hz. At this frequency, the sustainable harvester with adjustable output voltage in Sect. 3.3.2 is able to generate 487.5 μW output power. This is achievable when $C_{r1} = C_{r2} = 330 \mu F$, $C_{max=39} \mu F$, $n = 4$, and the output voltage is 10 V.

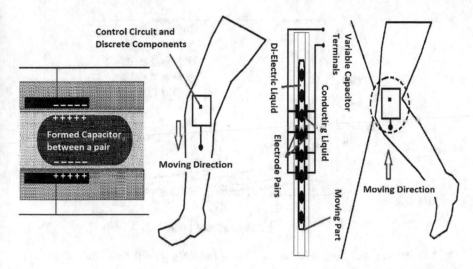

Fig. 3.48 An implementation of electrostatic harvester in [3] for energy harvesting from knee joint movement using the variable capacitor in [6]

According to Fig. 3.32b, the net generated energy of this harvester increases for higher values of n. However, the maximal energy efficiency is at $n = 3.4$. The value of n is considered to be 4 at which point the energy efficiency is 0.33 (very close to optimal efficiency) and the net generated energy is more than the case of $n = 3.4$. To calculate the output power, the expression in (3.147) is divided by $T_V = 1$ s. However, the variable capacitor that is used for this harvesting system is able to increase the frequency of changes in C_V (or decreasing T_V). Figure 3.49 shows the output power of this system for different values of C_{max} versus $1/T_V$. Each line in this figure is plotted for a specific output power. The more the frequency of changes in C_V, the less maximal capacitance is required for generating a specific output power. Using a variable capacitor with a maximal capacitance of $39\,\mu$F whose capacitance changes with a frequency of $20\,$Hz when it is linked to knee joint, this harvester generates $9.75\,$mW output power.

3.5.3 Energy Harvesting from Diaphragm Muscle

Diaphragm muscle as a low frequency and long amplitude energy sources in human body is discussed in this section. Later, energy harvesting from diaphragm muscle as an internal organ with considerable potential in powering a wide range of implantable medical devices is discussed.

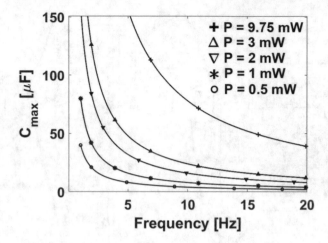

Fig. 3.49 The required C_{max} versus the movement frequency for different values of generated power

Fig. 3.50 The diaphragm muscle, its role in breathing and phases of its movement

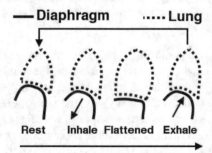

3.5.3.1 Diaphragm Muscle

The diaphragm is a strong muscle that plays a major role in breathing. Figure 3.50 shows a breathing cycle and the movement of this muscle. The air volume in lungs is minimal when the diaphragm muscle is at its rest position. During the inhale phase, this muscle moves downwards and the air is pumped into lungs to compensate the resulted vacuumed volume inside larger lungs. The duration of this phase depends on how deep inhalation is done. In this figure, a flattened shape of diaphragm denotes the maximum range that this muscle can move. During the exhale phase, the diaphragm moves upwards and the air is pushed out, until it reaches to its rest position. Depending on how deep the inhalation is done, the diaphragm muscle moves few to several centimetres during this cycle. For instance, it moves in a larger range when a person exercises.

Diaphragm muscle is a promising source of energy for providing an increasing number of implantable medical devices with a sufficient amount of energy. A permanent implanted energy harvester for energy harvesting from this muscle may generate sufficient energy for many medical applications continuously. This is contrary to energy harvesting from limbs, where the harvester generates energy only when a person moves. The amount of energy that is harvested from this muscle should not have any noticeable impact on the normal operation of this muscle, although it is a strong muscle. However, there seems not to be a strict limit on the size of a harvester for this acceptable amount of energy, since a space of several cubic centimetres is available (up to $480 \, cm^3$ [7]) around this muscle. Designing an efficient harvester for energy harvesting from sources of energy with a wide range of movement frequencies is challenging and sub-optimal. However, the movement frequency of diaphragm is relatively fixed and predictable in normal breathing. The movement characteristics of this muscle, i.e. low frequency and large amplitude, should be considered in designing an efficient harvester.

Piezoelectric harvesters are mostly used in the resonance mode to have an acceptable efficiency. This resonance frequency is not compatible with the low frequency of the diaphragm movement. The generated energy of electromagnetic harvesters decreases dramatically when the movement frequency is low [8]. However, the generated energy of electrostatic harvesters changes linearly with frequency. Structure of a variable capacitor that is capable of increasing the frequency of low frequency linear movements is explained in Sect. 3.5.1. Although it is possible to increase the frequency of the rotor of an electromagnetic harvester, but this is normally achievable with more sophisticated mechanical couplings compared to the structure of variable capacitors. Therefore, the variable-capacitance harvesters are notable candidates in energy harvesting from low frequency large amplitude kinetic energy sources.

3.5.3.2 Harvester and Simulations

The harvester in Sect. 3.3.3 is used for energy harvesting from diaphragm. In this harvester, a rechargeable battery provides the load with the required power and the core harvester charges the battery. In Sect. 3.3.3.3, it is shown both the net generated energy and the energy efficiency of this harvester increase for higher values of n. Therefore, the variable capacitor in Fig. 3.47 is selected with a maximal capacitance of $39 \, \mu F$ and a minimal capacitance of $31 \, nF$ in [4]. It is considered that this variable capacitor is able to increase the movement frequency with a ratio of 500 in [4]: $T_V = 3 \, ms$. The constant capacitors are chosen to be $330 \, \mu F$. Assuming the diaphragm's movement frequency equal to $0.34 \, Hz$, $V_B = 10 \, V$, this harvester is estimated to generate $298 \, mW$ power. This value is calculated by dividing the expression in (3.187) by $T_V = 3 \, ms$.

Figure 3.51 shows the simulation results of the harvester in Fig. 3.45 with the mentioned component values. This simulation is run for two sets of components: 1-ideal switches, 2-discrete SMD transistors, and diodes. The discrete SMD components are specified in Fig. 3.45. In this simulation, $R_L = 1 \, k\Omega$. The power delivered to R_L is constant, both for the ideal and discrete SMD switches. The total power delivered to the battery and R_L is $302 \, mW$ and is close to the calculated power of $298 \, mW$, when ideal switches are used. However, the total power delivered to the

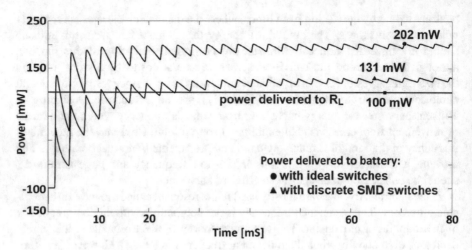

Fig. 3.51 The simulation of the harvester in Fig. 3.45 when it is used in energy harvesting from diaphragm muscle

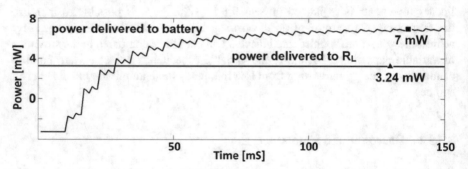

Fig. 3.52 The simulation of the harvester in Fig. 3.46 when it is used in energy harvesting from diaphragm muscle

battery and R_L reduces to 230 mW, when discrete SMD transistors and diodes are used. This reduction in the generated power is due to the conduction losses regarding the forward voltage drop across the diodes. However, the conduction losses of the discrete transistors are almost the same as when ideal switches were used, since the on for long enough. The reason for this has been detailed in Sect. 3.1.4.

Using the series diodes is necessary when discrete transistors are used, due to the intrinsic diodes between Drain–Source terminals of these transistors. These diodes are not required, when the core harvester is implemented using integrated transistors. To examine the feasibility of implementing the integrated version of the harvester, the harvester in Fig. 3.46 is simulated in the 0.18 µm technology. In this simulation, the switches and the control circuit are assumed to be on-chip, while C_{r1}, C_{r2}, and C_V are off-chip. Considering the same conditions for the energy source and the variable capacitor as for the discrete implementaion, Fig. 3.52 shows the simulation results, when the value of V_B is decreased to 1.8 V. This way, the

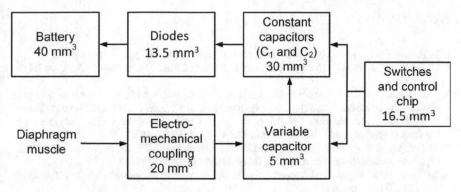

Fig. 3.53 The volume estimation presented in [4] for a discrete implementation of the system for energy harvesting from the diaphragm muscle

transistors in the 0.18 μm technology can tolerate the voltages across their terminals. Other parameters of the circuit, that is, C_{r1} and C_{r2} (off-chip capacitors), R_L, and T_V, are assumed to be the same as the discrete implementation. It is expected that the simulation result would be close to the calculated value from (3.187), since the series diodes are eliminated in the integrated implementation. The output power is calculated to be 10.5 mW with these component values. The total power delivered to the battery and R_L is 10.24 mW according to Fig. 3.52. This result is in close agreement with the calculated value, as expected.

Considerations regarding implementing a control block are reviewed in Sect. 3.4.3. The control circuit for these two harvesting systems is not very complex, and the switching frequency is relatively low. Therefore, a control circuit with a power consumption less than 10 μW is feasible. However, to make sure that the power consumption of the control block is not underestimated, a power consumption of 100 μW may be considered for this block. This amount of power is still much less than the amount of power generated by these two harvesting systems and, therefore, is not of any concern.

3.5.4   Volume Estimation

Figure 3.53 shows a volume estimation of the systems for energy harvesting from the knee joint movement and the diaphragm muscle in Sects. 3.5.2 and 3.5.3. In this figure, off-the-shelf discrete components are considered for implementing these electrostatic harvesting systems. This is due to the fact that components such as the constant capacitors, the variable capacitor, and the battery are off-the-shelf. However, it should be noted that integrated implementation of the control circuit and the transistors in the core harvester improves the performance. Overall, these harvesting systems fit in a volume less than 125 mm^3. With such a small volume, this device can be placed in the available space around the diaphragm muscle.

References

1. B. Razavi, *Design of Analog CMOS Integrated Circuits* (McGraw-Hill, Boston, 2001)
2. G. Gilman, Electrostatic Energy Generators and Uses of Same, U. S. Patent 6,936,994 B1, 30 Aug 2005
3. S.H. Daneshvar, M. Maymandi-Nejad, M. Sachdev, J. Redouté, A charge-depletion study of an electrostatic generator with adjustable output voltage. IEEE Sens. J. **19**(3), 1028–1039 (2019)
4. S.H. Daneshvar, M. Maymandi-Nejad, A new electro-static micro-generator for energy harvesting from diaphragm muscle. Int. J. Circuit Theory Appl. **45**(12), 2307–2328 (2017)
5. A. Kempitiya, D.A. Borca-Tasciuc, M.M. Hella, Low-power ASIC for microwatt electrostatic energy harvesters. IEEE Trans. Ind. Electron. **60**(12), 5639–5647 (2013)
6. T.N. Krupenkin, Method and Apparatus for Energy Harvesting Using Microfluidics, U. S. Patent 8,053,914 B1, 8 Nov 2011
7. T. Mussivand, P.J. Hendry, R.G. Masters, W.J. Keon, Progress with the heartSaver ventricular assist device. Ann. Thoracic Surg. 1999, **68**(2), 785–789 (1999)
8. J. Boland, Micro electret power generators. PhD. Thesis, California Intitute of Technology, Available online at: https://thesis.library.caltech.edu/5228/1/JustinBoland.pdf

Chapter 4
Asynchronous Electrostatic Harvesters

Abstract During a full energy conversion cycle, the capacitance of a variable capacitor changes from maximal to minimal and returns back to maximal again. The operation of a switching electrostatic harvester consists of four phases, in this period. At the start of consecutive investment and harvesting phases, a switching event should occur to initiate these phases. Similarly, another switching event should occur at the start of consecutive reimbursement and recovery phases to initiate these phases. The switching electrostatic harvesters are categorized to synchronous and asynchronous, based on whether or not these switching events are synchronized with the moments that the capacitance of the variable capacitor is maximal or minimal. In this chapter, fundamentals, different structures, and analysis of an asynchronous harvester with flyback circuit are covered.

4.1 Diode-Based Charge Transfer

In this section, the charge transfer between a variable capacitor and a storage component (a battery or a large capacitor) is discussed where a diode is placed between them. A diode connects these components at the moment that the voltage across them is almost the same. The impact of the series resistance, the type of the storage component on the conduction losses, and the shape of the current are detailed in this section.

4.1.1 Using a Diode

The energy source changes the capacitance of the variable capacitor during the investment and the harvesting phases. Subsequently, the voltage across the variable capacitor changes during these phases provided that firstly, the variable capacitor is initially charged before the start of these phases and secondly, the variable capacitor is isolated from the rest of the circuit (the charge in it is constant). The formula $Q = CV$ validates this occurrence. Placing a diode between the variable capacitor

S. H. Daneshvar et al., *Design of Miniaturized Variable-Capacitance Electrostatic Energy Harvesters*, https://doi.org/10.1007/978-3-030-90252-0_4

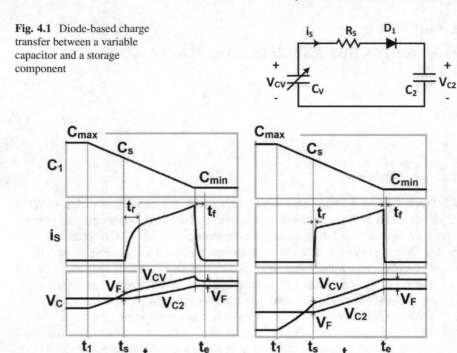

Fig. 4.1 Diode-based charge transfer between a variable capacitor and a storage component

Fig. 4.2 The impact of series resistance in the conducting path in diode-based charge transfer between a variable capacitor and a storage component that is not a battery or a large capacitor. (a) $t_r \lll t_e - t_s$. (b) $t_r \ll t_e - t_s$

and a storage component is widely used in many electrostatic harvesters to connect these two components autonomously.

Figure 4.1 shows the circuit representation of this type of charge transfer between the variable capacitor and a storage component. The direction of i_S in this figure is specified with the assumption that the voltage across the variable capacitor exceeds $V_F + V_{C2}$ (where V_F is the forward voltage drop across D_1) and an amount charge transfer from C_V to the storage component. Therefore, this figure represents when the voltage across C_V is increasing (the harvesting phase). R_S is the equivalent series resistance of C_V, C_2, and D_1 when D_1 is conducting.

Figure 4.2 shows the voltages across C_V and C_2 and the current that goes through these components (i_S), when the capacitance of C_V changes from a maximal to minimal value in Fig. 4.1. The capacitance of C_V is maximal and the voltage across C_V is less than the voltage across C_2 at the start: D_1 is off. The capacitance of C_V starts decreasing at $t = t_1$ and the voltage across C_V starts increasing since D_1 is off. At $t = t_s$, when the voltage across C_V exceeds $V_2 + V_F$ (and the capacitance of C_V is C_s), D_1 turns on and i_S should jump to a positive value. However, i_S could not jump abruptly to this positive value due to the time constant of the conducting path.

A rise time (t_r) is specified in Fig. 4.2: t_r is the period during which i_S increases from zero to the final positive value that is marked on this figure. This rise time depends on the resistance and the capacitance of the conducting path and is equal to:

$$t_r \approx 5\tau_s = 5R_SC_{ts}, \qquad C_{ts} = \frac{C_sC_2}{C_s + C_2}. \qquad (4.1)$$

The time constant of the conducting path is changing during the charge transfer period since; the capacitance of C_V changes during this time. D_1 continues conducting as long as C_V is decreasing; therefore, C_V and C_2 are connected until $t = t_e - t_f$. At this moment, the capacitance of C_V is minimal, $C_V = C_{min}$, and i_S should jump to zero. However, i_S reaches zero after a fall time equal to t_f at $t = t_e$. This fall time depends on the time constant of the conducting path at this moment, hence:

$$t_f \approx 5\tau_e = 5R_SC_{te}, \qquad C_{te} = \frac{C_{min}C_2}{C_{min} + C_2}. \qquad (4.2)$$

At $t = t_e$, the current that goes through C_V and C_2 is zero and the voltage difference between these capacitors reaches V_F (the forward voltage drop across D_1). The value of R_S changes the shape of i_S during the time that D_1 conducts. This impact is depicted for the cases that t_r is not negligible and is negligible compared to $t_e - t_s$ in Fig. 4.2a and b, respectively. Nonetheless, the integral of i_S in the time period of $t_s < t < t_e$ is independent of R_S, provided that i_S reaches zero at $t = t_e$. In the following subsections, the integral of i_S is calculated for different conditions of the storage component.

4.1.2 The Storage Is Not a Battery or a Large Capacitor

Considering the circuit in Fig. 4.1, the storage is not a large capacitor if the following condition is satisfied:

$$C_2 \gg C_{max}. \qquad (4.3)$$

Figure 4.2 shows the voltages across the variable capacitor and C_2 for this case. As can be seen, the voltages across C_V and C_2 are increasing during the charge transfer period. Therefore, these voltages are different at the end of the charge transfer period compared to the start of this period. Based on this figure, assuming that the voltage across C_V is V_1 at $t = t_s$, the voltage across C_2 is $V_2 = V_1 - V_F$ at this moment. The current is still zero at this moment, and the charge in both C_V and C_2 is

$$q_1 = C_sV_1 + C_2V_2 = C_sV_1 + C_2(V_1 - V_F). \qquad (4.4)$$

At the end of the charge transfer period, $t = t_e$, when the capacitance of the variable capacitor is minimal, i_S becomes zero again and the charge in both C_V and C_2 is obtained as follows:

$$q_f = C_{min} V_{1f} + C_2(V_{1f} - V_F).$$ (4.5)

Since the current is zero at the start and at the end of the charge transfer period, q_1 and q_f in the above expressions are equal, according to the charge conservation law. Equating q_1 and q_{1f}, V_{1f} is obtained as follows:

$$V_{1f} = \frac{C_s + C_2}{C_{min} + C_2} V_1.$$ (4.6)

The current at the end of this period is zero and the voltage difference between C_V and C_2 equals to V_F, as depicted in Fig. 4.2. Therefore, the voltage across C_2 at end of this charge transfer period is:

$$V_{2f} = V_{1f} - V_F,$$ (4.7)

where V_{1f} is obtained in (4.6). The voltage across C_2 increases from $V_1 - V_F$ to $V_{1f} - V_F$ during this charge transfer period; therefore, the integral of i_S is calculated as follows:

$$V_{2f} = V_2 + \frac{1}{C_2} \int i_S dt$$

$$\Rightarrow V_{1f} - V_F = V_1 - V_F + \frac{1}{C_2} \int i_S dt$$

$$\Rightarrow \int i_S dt = C_2 \left(V_{1f} - V_1\right) = C_2 \frac{C_s - C_{min}}{C_{min} + C_2} V_1.$$ (4.8)

As can be seen in the above expression, the integral of i_S is not dependent on the value of R_S. Therefore, the integral of i_S is the same for any value of R_S, as long as C_V connects to C_2 when $C_V = C_s$, and $V_{C_V} = V_1$.

4.1.3   The Storage Is a Battery or a Large Capacitor

In this case, the storage component is a battery or a large capacitor with the following condition being satisfied:

$$C_2 \gg C_{max}.$$ (4.9)

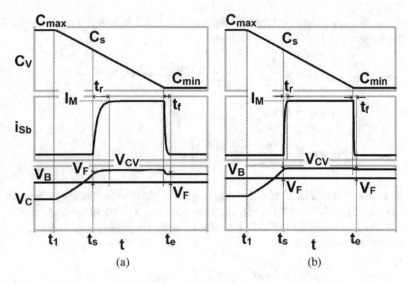

(a)   (b)

Fig. 4.3 The impact of series resistance in the conducting path in diode-based charge transfer between a variable capacitor and a battery (or a large capacitor). (a) $t_r \ll t_e - t_s$. (b) $t_r \ll t_e - t_s$

Figure 4.3 shows the series current (i_{Sb}) and the voltages across the variable capacitor and the battery (or the large capacitor) during a charge transfer period. These signals are plotted for different values of R_S in Fig. 4.3a and b. In Fig. 4.3b, the value of R_S is chosen to make the time constant of the conducting path negligible compared to the duration that D_1 conducts. However, the time constant of the conducting path is not negligible compared to the duration that D_1 conducts in Fig. 4.3a for a bigger value of R_S.

In Fig. 4.2, the current is increasing during the charge transfer period, since the voltages across C_1 and C_2 are increasing during this time. However, the voltage across the storage component does not change when it is a battery or a large capacitor. Accordingly, the voltage across the variable capacitor at the end of this period is the same as it is at the start of this period. Therefore, as depicted in Fig. 4.3, the series current reaches a positive value at $t = t_s + t_r$ and remains at this value for the whole charge transfer period. This positive current is marked as I_M in Fig. 4.3 and:

$$i_{Sb}(t) = I_M, \qquad t_s + t_r < t < t_e - t_f. \qquad (4.10)$$

Considering that the voltage across the storage component is equal to V_B, D_1 starts conducing when the voltage across C_V is equal to $V_B + V_F$ at $t = t_s$. At $t = t_s + t_r$, the voltage across the variable capacitor reaches the following value and remains at this value until $t = t_e - t_f$:

$$V_{CV} = V_B + V_F + R_S I_M, \qquad t_s + t_r < t < t_e - t_f, \qquad (4.11)$$

where I_M is the constant current during this time and is marked on Fig. 4.3. This current is obtained as follows:

$$i_{Cv} = C\frac{dV}{dt} + V\frac{dC}{dt} \xrightarrow{V \text{ is constant}} i_{Cv} = V\frac{dC}{dt}$$

$$\Rightarrow I_M = (V_B + V_F + R_S I_M)\left|\frac{dC}{dt}\right|$$

$$\Rightarrow I_M = \frac{V_B + V_F}{1 - R_S\left|\frac{dC}{dt}\right|}\left|\frac{dC}{dt}\right|. \tag{4.12}$$

The above expression is written assuming the positive value for i_{Sb} with specified direction in Fig. 4.1. Therefore, the absolute value of dC/dt is used. According to Fig. 4.3, the absolute value of dC/dt is constant for the whole duration of $t_s < t < t_e - t_f$ and is equal to:

$$\left|\frac{dC}{dt}\right| = \frac{C_{max} - C_{min}}{t_e - t_f - t_s}. \tag{4.13}$$

Using above expressions, the value of I_M is calculated. In (4.8), the integral of i_S is calculated for the case that C_2 is not much larger than C_{max}. Using this expression, the integral of i_{Sb} when $C_2 \gg C_{max}$ is found as following:

$$\int i_S dt = C_2\frac{C_s - C_{min}}{C_{min} + C_2}V_1 \xrightarrow{for\ i_{Sb}:\ C_2 \gg C_{min}} \int i_{Sb} dt = C_2\frac{C_s - C_{min}}{C_2}V_1, \tag{4.14}$$

where V_1 is the voltage across C_V at the beginning of the charge transfer period. Therefore:

$$\int i_{Sb} dt = (C_s - C_{min})(V_B + V_F). \tag{4.15}$$

All the values for I_M, t_r, and t_f depend on R_S, and the impact of R_S on the shape of i_S is illustrated in Fig. 4.3. However, the integral of i_S does not depend on R_S according to the above expression.

4.1.4 Energy Transfer and Conduction Losses

Figure 4.4 shows the energy transfer between C_V and C_2 and the conduction losses in R_S and D_1 of the circuit in Fig. 4.1. The energy transfer between two constant capacitors occurs due to the voltage difference between the capacitors, as depicted in Fig. 3.6. However, the energy transfer between C_V and C_2 (with the scenario that is explained in this section) occurs as a result of an energy source changing the

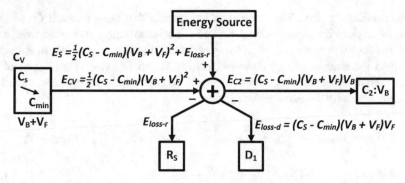

Fig. 4.4 The energy transfer between C_V and C_2 and the conduction losses in R_S and D_1 of the circuit in Fig. 4.1

capacitance of C_V. Therefore, an energy source block is added to Fig. 4.4, compared to Fig. 3.6. An energy expression is shown over each line. The steps to derive these expressions are detailed in the following paragraphs.

The voltage across the variable capacitor at the start and at the end of the charge transfer period is $V_B + V_F$. However, the capacitance of the variable capacitor decreases to C_{min} from C_s during this period. A downward arrow in Fig. 4.4 depicts this change in the capacitance of C_V. The amount of energy that is transferred from C_V at the end of the charge transfer period is:

$$E_{CV} = \frac{1}{2} \left(C_s - C_{min} \right) \left(V_B + V_F \right)^2. \tag{4.16}$$

The voltage across the storage component (a battery or a large capacitor in this case) is constant and equal to V_B during the charge transfer period. The energy that is received by this storage component is:

$$E_{C2} = V_B \int i_{Sb} dt = \left(C_s - C_{min} \right) \left(V_B + V_F \right) V_B, \tag{4.17}$$

where integral of i_{Sb} is calculated in (4.15). The conduction losses in D_1 are calculated as below:

$$E_{loss-d} = V_F \int i_{Sb} dt = V_F \left(C_s - C_{min} \right) \left(V_B + V_F \right). \tag{4.18}$$

As can be seen in the above expressions, the sum of the received energy in C_2 (E_{C2}) and the lost energy in D_1 is more than the energy that is obtained from C_V (E_{CV}). The energy source decreases the capacitance of the variable capacitor while the voltage across it is constant. Therefore, the charge in the variable capacitor decreases according to $Q = CV$. The changes in the capacitance of C_V make this charge transfer possible. An amount of energy is also lost due to existence of R_S

in the conducting path. The integral of the current that goes through C_2 and C_V does not depend on R_S according to (4.15). Therefore, the conduction loss in R_S increases for higher values of R_S. However, this higher conduction loss in R_S would not affect the amount energy changes in C_V, C_2, and D_1, since the integral of the series current is the same as for any other R_S. Therefore, the conduction loss in R_S is compensated by the energy source and:

$$E_S - E_{loss-r} = E_{C2} + E_{loss-d} - E_{CV} = \frac{1}{2}(C_s - C_{min})(V_B + V_F)^2$$

$$\Rightarrow E_S = \frac{1}{2}(C_s - C_{min})(V_B + V_F)^2 + E_{loss-r}, \tag{4.19}$$

where E_S is the amount of energy that is added by the energy source. According to the above expression, the difference between E_S and E_{loss-r} is not dependent on R_S. To have a better understanding of this relation, E_{loss-r} is calculated as follows for $R_S = 0$.

When $R_S = 0$, the current that goes through C_V and the storage component is similar to Fig. 4.3b with t_r, $t_f \approx 0$. The value of I_M in this figure is calculated in (4.12). In (3.21), E_{loss-r} is not equal to zero for $R_S = 0$, since i_S has an impulse at $t = t_s$ for this value of R_S. However, i_{Sb} here has a rectangular shape for $R_S = 0$ and i_S does not have any impulses during the charge transfer period. Therefore, writing E_{loss-r} as in (3.22), the conduction losses are zero for $R_S = 0$ here. In this case, the energy that is added to the system by the energy source is:

$$E_S = \frac{1}{2}(C_s - C_{min})(V_B + V_F)^2. \tag{4.20}$$

The conduction loss in R_S depends on R_S: The higher the value of R_S, the higher the conduction loss in this resistance is. However, the conduction loss in D_1 (E_{loss-d}), E_{CV}, and E_{C2} does not depend on R_S. Therefore, E_S increases and compensates for the higher resistive conduction loss, when the resistance of the conducting path is higher. This way, the conduction loss in D_1, the energy that is obtained from C_V, and the energy that is received by C_2 do not change for different values of R_S.

4.2   Non-sustainable Core Asynchronous Harvester

A non-sustainable asynchronous electrostatic harvester is studied in this section. This core harvester accumulates the harvested energy in a storage capacitor. The harvester stops generating net energy when the voltage across the storage capacitor saturates to its final value. Adding a flyback circuit to this core asynchronous harvester is discussed in the next section to resolve this issue.

Fig. 4.5 The core
asynchronous electrostatic
harvester

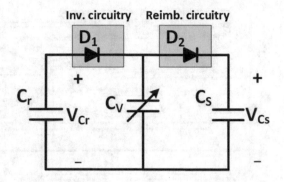

4.2.1   Operation and the Voltages Across the Capacitors

Figure 4.5 shows the core asynchronous electrostatic harvester. The investment
and the reimbursement circuitries of this harvester are highlighted in this figure.
The investment circuitry of this harvester is a single diode (D_1) that is placed
between the reservoir capacitor, C_r, and the variable capacitor, C_V. Similarly, the
reimbursement circuitry is a single diode (D_2) that is placed between C_V and the
storage component, C_S. The diode, D_1, starts conducting when the voltages across
C_V and C_r become equal. D_2 starts the reimbursement phase when the voltages
across C_V and C_S become equal. These diodes initiate the investment and the
reimbursement phases autonomously. The start of these phases are not synchronized
with the moments that the capacitance of C_V is maximal or minimal. Therefore,
this harvester is asynchronous. During the investment phase, an amount of charge
is obtained from C_r. However, no charge is transferred back to this capacitor later;
therefore, the harvester is non-sustainable.

The voltages across the capacitors of this harvester are shown in Fig. 4.6. In
this figure, the phases of operation are specified at the kth cycle, and the voltage
across C_S is marked as $V_{CS}(k)$ at the start of the investment phase of this cycle. It is
assumed that C_r is much bigger than the other capacitors in this harvester. Therefore,
the changes in the voltage of this capacitor are considered to be negligible. This
capacitor is charged to V_{ri} initially. The operation of this harvester is explained as
follows according to these figures.

As can be seen in Fig. 4.6, the investment and the recovery phases overlap in this
harvester. The voltage across C_V decreases during the recovery phase, since it is
isolated from the rest of the circuit: D_1 and D_2 are off. The investment phase starts
during the recovery phase when the voltage across C_V becomes less than the voltage
across C_r by V_F (the forward voltage drop across D_1). At this moment, D_1 turns
on and C_r connects to C_V. During the recovery phase, C_r keeps the voltage across
C_V constant at $V_{ri} - V_F$, since $C_r \gg C_V$. The investment phase ends when the
capacitance of C_V reaches its maximal value, C_{max}.

The energy source starts changing the capacitance of C_V from its maximal value
towards its minimal value: The harvesting phase starts at this moment. The variable

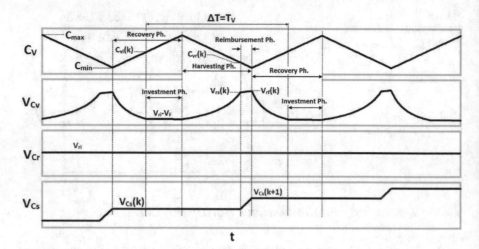

Fig. 4.6 The phases of operation and the voltages across the capacitors of the core asynchronous harvester in Fig. 4.5

capacitor is isolated from the rest of the circuit during this phase, until the voltage across it reaches the following value:

$$V_{rs}(k) = V_{Cs}(k) + V_F, \tag{4.21}$$

where the term "rs" denotes the voltage across C_V at the start of the reimbursement phase, and it is assumed the forward voltage drop across D_2 is the same as D_1. The voltage across C_S is $V_{Cs}(k)$ at the start of the investment phase. This capacitor is isolated from the rest of the circuit (D_2 is off) until the start of the reimbursement phase. Therefore in the above expression, the voltage across C_S at the start of the reimbursement phase is equal to $V_{Cs}(k)$. This can be seen in Fig. 4.6. The capacitance of the variable capacitor at the start of the reimbursement phase ($C_{vr}(k)$) is obtained as follows, based on the charge conservation law in C_V:

$$C_{max}(V_{ri} - V_F) = C_{vr}(k)(V_{Cs}(k) + V_F) \Rightarrow C_{vr}(k) = \frac{V_{ri} - V_F}{V_{Cs}(k) + V_F}C_{max}. \tag{4.22}$$

The voltage across C_V at the end of the reimbursement phase is obtained as follows, based on Sect. 4.1.2:

$$V_{rf}(k) = \frac{C_{vr}(k) + C_S}{C_{min} + C_S}(V_{Cs}(k) + V_F), \tag{4.23}$$

where the term "rf" denotes the end of the reimbursement phase and C_{vr} is found in (4.22). The energy source changes the capacitance of C_V from its minimal value towards its maximal value, and the recovery phase starts at this moment. The voltage across C_V decreases while it is isolated from the rest of the circuit, until it reaches a voltage less than the voltage across C_r. The investment phase of the $(k + 1)$th cycle starts at the moment that the voltage across C_V reaches $V_{ri} - V_F$. Therefore, the capacitance of C_V at the start of the investment phase of the $(k + 1)$th is found as follows:

$$C_{vi}(k + 1) = \frac{V_{rf}(k)}{V_{ri} - V_F} C_{min}.$$ (4.24)

The voltage across C_S at the start of the investment phase of the $(k + 1)$th cycle is equal to this voltage at end of the reimbursement phase of the kth cycle. Therefore:

$$V_{Cs}(k + 1) = V_{rf}(k) - V_F$$

$$\Rightarrow V_{Cs}(k + 1) = \frac{C_{max}(V_{ri} - V_F) + C_S(V_{Cs}(k) + V_F)}{C_{min} + C_S} - V_F$$

$$\Rightarrow V_{Cs}(k + 1) = \frac{C_{min}\left(n(V_{ri} - V_F) - V_F\right) + C_S V_{Cs}(k)}{C_{min} + C_S},$$ (4.25)

where $C_{vr}(k)$ in (4.23) is replaced with its value in (4.22). The above expression may be rewritten as follows:

$$V_{Cs}(k + 1) - \frac{C_S}{C_S + C_{min}} V_{Cs}(k) = \frac{C_{min}}{C_S + C_{min}}\left(n(V_{ri} - V_F) - V_F\right).$$ (4.26)

The above recursive formula is linear and non-homogeneous. This formula could be solved as following based on Sect. 2.4.2:

$$V_{Cs}(k) = V_{Cs-g}(k) + V_{Cs-p}(k)$$

$$\Rightarrow V_{Cs}(k) = d_1\left(\frac{C_S}{C_S + C_{min}}\right)^k + \frac{\frac{C_{min}}{C_S + C_{min}}\left(n(V_{ri} - V_F) - V_F\right)}{1 - \frac{C_S}{C_S + C_{min}}}$$

$$\Rightarrow V_{Cs}(k) = d_1\left(\frac{C_S}{C_S + C_{min}}\right)^k + n(V_{ri} - V_F) - V_F,$$ (4.27)

where d_1 is a constant coefficient. To find this coefficient, the initial conditions (initial voltages across the capacitors) should be used. Considering that C_r is initially charged to V_{ri}, the voltages across C_V and C_S are $V_{ri} - V_F$ and $V_{ri} - 2V_F$ respectively at the beginning of the harvester operation. The initial voltages across the capacitors of this harvester are related as this, due to the existence of D_1 and D_2 between these capacitors, therefore:

$$V_{Cs}(0) = V_{ri} - 2V_F$$

$$\Rightarrow V_{Cs}(0) = d_1 + n(V_{ri} - V_F) - V_F$$

$$\Rightarrow V_{Cs}(k) = n(V_{ri} - V_F) - V_F - (V_{ri} - V_F)(n - 1)\left(\frac{C_S}{C_S + C_{min}}\right)^k, \quad (4.28)$$

where $k = 0$ is replaced in (4.27) to find d_1.

4.2.2 The Saturation Issue

According to (4.28), the voltage across C_S eventually saturates to the following value (V_{Cs-f}), since the term that is powered to k is less than 1:

$$V_{Cs-f} = \lim_{k \to \infty} V_{Cs}(k) = n(V_{ri} - V_F) - V_F. \quad (4.29)$$

The voltage across C_V at the start of the reimbursement phase would be $n(V_{ri} - V_F)$ according to (4.21), when the voltage across C_S saturates to the above value. The voltage across C_V changes between $V_{ri} - V_F$ and $n(V_{ri} - V_F)$ during the harvesting phase at this time. Therefore, no energy transfers between C_r, C_V, and C_S during the investment and the reimbursement phases and the net generated energy in the harvester are zero at this time: The harvester stops working properly. This can also be validated by replacing the above value in (4.22) and (4.24). According to these expressions, the capacitance of C_V at the start of the investment and the reimbursement phases are C_{max} and C_{min}, respectively, when the voltage across C_S saturates to its final value. Therefore, the duration of these phases reaches zero at this time.

4.2.3 Simulation and QV Diagram

The simulation results of this harvester are shown with solid lines in Fig. 4.7. The component values for this simulation are $C_r = 10\,\mu F$, $C_S = 10\,nF$, $C_{min} = 1\,nF$, $n = 5$, and $V_F = 0.5\,V$. The initial voltage across C_r is 5 V and the voltage decrease across this capacitor is negligible, since $C_r \gg C_V, C_S$. The scattered circles in this figure show the obtained discrete values from the derived expressions for $V_{Cs}(k)$ and $V_{rs}(k)$ in (4.28) and (4.21). These circles are connected with dashed lines to make the comparison between theoretical and simulation results easier. As can be seen, the theoretical and simulation results are matched.

The switching events in this core harvester are asynchronous: The start of the investment and the reimbursement phases do not need to be synchronized with the moments that $C_V = C_{max}$ and $C_V = C_{min}$. The investment and the reimbursement circuitries are diodes; therefore, this core harvester does not need any control circuit. However, the variable capacitor does not connect to C_r and C_S during the investment and the reimbursement phases when the voltage across C_S reaches its final value.

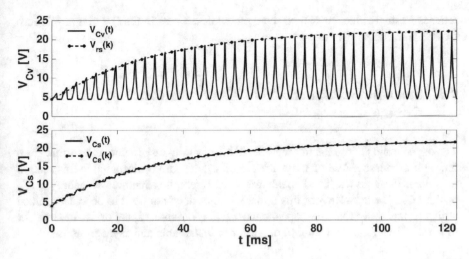

Fig. 4.7 Comparison of simulation results with the derived expressions for the harvester in Fig. 4.5

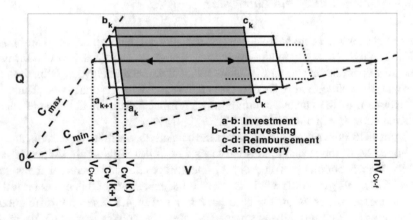

Fig. 4.8 The QV diagram of C_V in the core asynchronous harvester of Fig. 4.5

Therefore, the deliverable energy and subsequently the net generated energy become zero. To illustrate this, the QV diagram of this harvester is shown in Fig. 4.8 for the case that $C_r \gg C_V, C_S$. The QV diagram is different in each cycle for this harvester. The QV diagram of the kth cycle and a few next cycles is shown in this figure. The voltage decrease across C_r is not negligible, since $C_r \gg C_V, C_S$. Therefore, the variable capacitor in each cycle is charged to a voltage less than the previous cycle at the end of the investment phase. The enclosed area in the QV diagram of the kth cycle is highlighted. As can be seen this area, which is equal to the deliverable energy, becomes less from one cycle to the next one. Skipping several QV diagrams, the line with two arrows on it shows the QV diagram, when the voltage across C_S

reaches its final value. As can be seen, the enclosed area in the QV diagram is zero from this moment onwards.

4.3   Sustainable Asynchronous Harvester with Flyback Circuit

The asynchronous core harvester in Sect. 4.2.1 is non-sustainable and suffers from the saturation issue. Adding a flyback circuit to this core is discussed in this section to address these issues. The flyback circuit transfers the accumulated energy in C_S back to C_r. The activation of this flyback circuit does not need to be synchronized with the moments that the capacitance of the variable capacitor is maximal or minimal. Therefore, the resulting harvester is sustainable and asynchronous.

4.3.1   Operation of the Harvester

Figure 4.9 shows the core harvester in Fig. 4.5 with a flyback circuit that is proposed in [1]. The flyback circuit is highlighted in this figure. The flyback circuit turns on every few operating cycles and transfers the accumulated energy in C_S back to C_r. This way, the voltage across C_S decreases before it reaches its final value. Therefore the operation of the harvester returns to its normal before the deliverable energy becomes zero, as explained in Sect. 4.2.1.

Figure 4.10 shows the voltages across the capacitors, the inductor current, and the current in D_1 of the asynchronous harvester in Fig. 4.9. This figure shows the case that the flyback circuit turns on every K_t operating cycles and decreases the voltage across C_S from V_2 to V_1. The voltages across the capacitors of this harvester follow the same expressions as for the core harvester in Fig. 4.5. The only difference is that every time that the flyback circuit activates, the voltage across C_S decreases

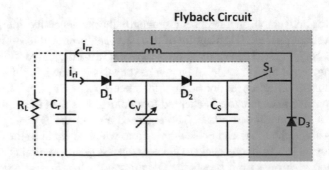

Fig. 4.9   The core harvester in Fig. 4.5 with a flyback circuit proposed in [1]

to a lower value. Using this value in (4.28), a relevant value is found such that $V_{Cs}(k_1) = V_1$. Next voltages across C_S could be found by putting values of $k > k_1$ until $k = k_2$ when the flyback circuit activates again. At the end of the k_2 cycle, the voltage across C_S is $V_{Cs}(k_2 + 1)$. This is evident from Fig. 4.6. During the time that the flyback circuit is activated, a current goes through the inductor and C_r. This current is shown in Fig. 4.10 and is detailed shortly in the following paragraphs to calculate the net generated energy in this harvester.

Based on Sect. 4.2.1, the voltage across C_V changes between $V_{ri} - V_F$ and $V_{rf}(k)$ in each cycle. According to (4.25), $V_{rf}(k)$ may be obtained based on $V_{Cs}(k)$: $V_{rf}(k) = V_{Cs}(k+1) + V_F$. Therefore, changes in the voltage across C_V in each cycle depend on the voltage across C_S. Adding the flyback circuit to the core harvester, the voltage across C_S may be controlled between a lower voltage limit, V_1, and an upper voltage limit, V_2, as shown in Fig. 4.11. This figure shows the voltage across

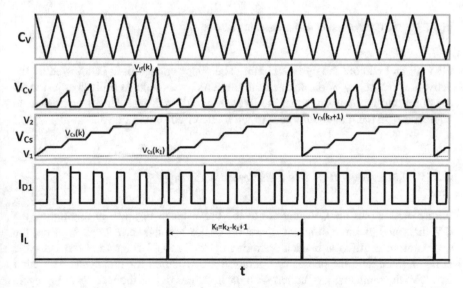

Fig. 4.10 The voltages across the capacitors and the inductor current of the asynchronous harvester in Fig. 4.9, when the flyback circuit turns on each six operating cycles

Fig. 4.11 The voltage across C_S in the harvester of Fig. 4.9 with and without the flyback circuit [2]

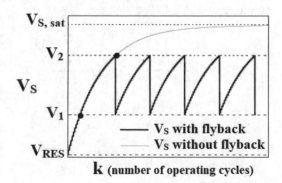

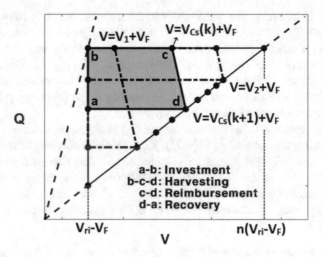

Fig. 4.12 The QV diagram of the harvester in Fig. 4.9

C_S with and without the flyback circuit. The upper voltage limit is adjusted by the activation frequency of the flyback circuit. The lower voltage limit is adjusted by the duration that S_1 is on, when the flyback circuit activates. This is equivalent to adjusting t_1 in the magnified figure of the inductor current in Fig. 5.2.

4.3.2 Calculation of the Deliverable Energy

Figure 4.12 shows the QV diagram of the harvester in Fig. 4.9. In this figure, the QV diagrams of three different operating cycles are shown. As can be seen, the QV diagram is different in each operating cycle. The QV diagram of the kth cycle is highlighted, and the phases of operation are specified for this cycle. The other two QV diagrams that are plotted with dashed lines depict the last operating cycle before the flyback circuit activates and the first operating cycle after the flyback circuit activates. The amount of deliverable energy in the kth cycle is equal to the greyed area in Fig. 4.12, therefore:

$$E_{del}(k) = \left(\frac{V_{Cs}(k+1) + V_{Cs}(k) + 4V_F - 2V_{ri}}{2} \right)$$

$$\left(C_{max}\left(V_{ri} - V_F \right) - C_{min}\left(V_{Cs}(k+1) + V_F \right) \right). \quad (4.30)$$

Using the expression in (4.28) for $V_{Cs}(k)$ and $V_{Cs}(k+1)$, the above formula is simplified as follows:

$$E_{del}(k) = \frac{1}{2}C_{min}(n-1)^2(V_{ri} - V_F)^2 r\left(2r^k - r^{2k+1} - r^{2k}\right), \qquad (4.31)$$

where:

$$r = \frac{C_S}{C_S + C_{min}}. \qquad (4.32)$$

The flyback circuit activates every K_t operating cycles and resets the voltage across C_S to its initial value. Therefore, the average deliverable energy in each cycle of this harvester is calculated as below:

$$\begin{aligned}
E_{del} &= \frac{1}{K_t}\sum_{k=k_1}^{k=k_2} E_{del}(k) = \frac{1}{2}C_{min}(n-1)^2(V_{ri}-V_F)^2 r\sum_{k=k_1}^{k=k_2}\left(2r^k - r^{2k+1} - r^{2k}\right) = \\
&\frac{1}{2K_t}C_{min}(n-1)^2(V_{ri}-V_F)^2 r\left(2\frac{r^{k_1} - r^{k_2+1}}{1-r} - (1+r)\frac{r^{2k_1} - r^{2(k_2+1)}}{1-r^2}\right) \\
&\frac{1}{2K_t}C_{min}(n-1)^2(V_{ri}-V_F)^2\frac{r}{1-r}\left(r^{k_1} - r^{k_2+1}\right)\left(2 - r^{k_1} - r^{k_2+1}\right),
\end{aligned} \qquad (4.33)$$

where k_1 and $k_2 + 1$ are related to the upper and lower voltage limits (V_1 and V_2) as below:

$$V_1 = V_{Cs}(k_1) = n(V_{ri} - V_F) - V_F - (V_{ri} - V_F)(n-1)r^{k_1},$$
$$V_2 = V_{Cs}(k_2 + 1) = n(V_{ri} - V_F) - V_F - (V_{ri} - V_F)(n-1)r^{k_2+1}. \qquad (4.34)$$

From the above expressions, the following results are obtained:

$$V_2 - V_1 = (n-1)(V_{ri} - V_F)\left(r^{k_1} - r^{k_2+1}\right)$$
$$V_2 + V_1 - 2(V_{ri} - 2V_F) = 2 - r^{k_1} - r^{k_2+1}, \qquad (4.35)$$

Using the above results in (4.33), the total deliverable energy before the flyback circuit activates is obtained as follows:

$$E_{del} = \frac{1}{2K_t}C_S\left(V_2 - V_1\right)\left(V_2 + V_1 - 2(V_{ri} - 2V_F)\right). \qquad (4.36)$$

4.3.3   Calculation of the Net Generated Energy

The accumulated energy in C_S transfers to C_r every time that the flyback circuit activates. Therefore, the net generated energy in C_S is zero. This is evident from Fig. 4.10, where the voltage across C_S increases from V_1 to V_2 and decreases back to V_1 when the flyback circuit activates. The net generated energy of this harvester is equal to the net generated energy in C_r. The expression in (2.28) may be used to find the net generated energy in C_r. Therefore, the integral of the investment and the reimbursement currents in this capacitor should be calculated.

In Sect. 4.2.1 and Fig. 4.6, the reimbursement phase is defined as the time that C_V connects to C_S through D_2. An amount of charge is transferred from C_V to C_S during this time, hence this duration is the reimbursement phase for C_S. However in the circuit of Fig. 4.9, the net generated energy in C_S is zero and the net generated energy of the harvester is equal to the net generated energy in C_r. Therefore, the reimbursement phase should be defined for C_r. The reimbursement phase for this capacitor is the time that the flyback circuit turns on. During the flyback circuit activation, C_S connects to C_V through S_1, D_3, and the inductor. An amount of charge is transferred from C_S to C_r during this time. This way, the flyback circuit activation is equivalent to reimbursement phase for C_r.

The inductor in the flyback circuit is selected to be large enough, so that the resistive conduction losses become negligible. This condition and the inductor current in this case are detailed in Sect. 5.1.2. Figure 5.2 in this section shows the magnified inductor current during the time that the flyback circuit is on. The operation of the flyback circuit is explained based on this figure as follows. The switch, S_1 of the flyback circuit in Fig. 4.9, turns on at $t = t_s$, when the voltage across C_S is equal or more than the higher voltage limit, V_2. After S_1 turns on, C_S connects to C_r through the inductor and the inductor current increases from zero. The inductor current follows a sinusoidal shape and may be expressed as follows:

$$i_{rr1}(t) = \frac{V_2 - V_{ri}}{L\omega_1} sin(\omega_1(t - t_s)), \quad t_s < t < t_1, \tag{4.37}$$

where $t = t_1$ is when the switch, S_1, turns off and V_2 is the upper voltage limit across C_S. The resistive conduction losses during this time are negligible, since a large value for the inductor is chosen to satisfy the condition in (5.1). In the above expression:

$$\omega_1 = \frac{1}{\sqrt{LC_t}}, \quad C_t = \frac{C_S C_r}{C_S + C_r} \xrightarrow{C_r \gg C_S} C_t \approx C_S. \tag{4.38}$$

Using the above expressions, the integral of $i_{rr1}(t)$ is calculated as below:

$$\int_{t_s}^{t_1} i_{rr1}(t)dt = C_S(V_2 - V_{ri})\Big(1 - cos(\omega_1 t_1)\Big). \tag{4.39}$$

This current goes through C_S, C_r, and the inductor. The voltage across C_S decreases to V_1 from V_2 during the time that S_1 is on. Therefore:

$$V_1 = V_2 - \frac{1}{C_S} \int_{t_s}^{t_1} i_{rr1}(t)dt \Rightarrow \int_{t_s}^{t_1} i_{rr1}(t)dt = C_S(V_2 - V_1). \qquad (4.40)$$

Using the expressions in (4.39) and (4.40):

$$C_S(V_2 - V_1) = C_S(V_2 - V_{ri})\left(1 - cos(\omega_1 t_1)\right)$$

$$\rightarrow 1 - cos(\omega_1 t_1) = \frac{V_2 - V_1}{V_2 - V_{ri}} \Rightarrow cos(\omega_1 t_1) = \frac{V_1 - V_{ri}}{V_2 - V_{ri}}$$

$$\Rightarrow sin^2(\omega_1 t_1) = \frac{(V_2 - V_1)(V_2 + V_1 - 2V_{ri})}{(V_2 - V_{ri})^2}. \qquad (4.41)$$

The above result will be used in calculating the integral of i_{rr} in the following paragraphs.

At $t = t_1$, S_1 turns off and the stored energy in the inductor transfers to C_r through D_3. This continues until the inductor current reaches zero at $t = t_0$. The inductor current during this time is expressed as follows:

$$i_{rr2}(t) = I_0 cos(\omega_2(t - t_1)) - \frac{V_{ri} + V_F}{L\omega_2} sin(\omega_2(t - t_1)), \quad t_1 < t < t_0, \qquad (4.42)$$

where I_0 is the inductor current at $t = t_1$ and can be found from (4.37) and:

$$I_0 = \frac{V_2 - V_{ri}}{L\omega_1} sin(\omega_1(t_1 - t_s)), \quad \omega_2 = \frac{1}{\sqrt{LC_r}}. \qquad (4.43)$$

The maximum value of $i_{rr2}(t)$ in (4.42) is as follows:

$$i_{rr2-max} = \sqrt{I_0^2 + \left(\frac{V_{ri} + V_F}{L\omega_2}\right)^2}. \qquad (4.44)$$

The above value is much bigger than $i_{rr2}(t_1)$. This can be shown using (4.43) as below:

$$I_0^2 = \frac{1}{L}C_S(V_2 - V_{ri})^2 sin^2(\omega_1 t_1), \quad \left(\frac{V_{ri} + V_F}{L\omega_2}\right)^2 = \frac{1}{L}C_r(V_{ri} + V_F)^2,$$

$$\xrightarrow{C_r \gg C_S} I_0^2 \ll \left(\frac{V_{ri} + V_F}{L\omega_2}\right)^2 \Rightarrow i_{rr2-max} \gg i_{rr2}(t_1) = I_0. \qquad (4.45)$$

According to the above relation, the inductor current during $t_1 < t < t_0$ may be assumed linear with high accuracy, therefore:

$$\int_{t_1}^{t_0} i_{rr2}(t)dt = \frac{LI_0^2}{2(V_{ri} + V_F)} = \frac{C_S(V_2 - V_{ri})^2}{2(V_{ri} + V_F)}sin^2(\omega_1 t_1). \tag{4.46}$$

Using the result in (4.41) in the above expression:

$$\int_{t_1}^{t_0} i_{rr2}(t)dt = \frac{C_S(V_2 - V_1)(V_2 + V_1 - 2V_{ri})}{2(V_{ri} + V_F)}. \tag{4.47}$$

The integral of i_{rr} is written as below, using the results in (4.40) and (4.47):

$$\int_{t_s}^{t_0} i_{rr}dt = \int_{t_s}^{t_1} i_{rr1}dt + \int_{t_1}^{t_0} i_{rr2}dt$$

$$\Rightarrow \int_{t_s}^{t_0} i_{rr}dt = \frac{C_S(V_2 - V_1)(V_2 + V_1 + 2V_F)}{2(V_{ri} + V_F)}. \tag{4.48}$$

Knowing the integral of i_{rr} as above, the integral of i_{ri} is found as follows to calculate E_{net}. During the investment phase of the kth cycle, the variable capacitor connects to C_r when its capacitance is equal to $C_{vi}(k)$. This phase continues until the capacitance of C_V reaches C_{max}. During this phase, the investment current, i_{ri}, goes through C_r and C_V. The voltage across C_V is kept constant at $V_{ri} - V_F$, since $C_r \gg C_V$. The integral of this current is detailed in Sect. 4.1.3 and is calculated in (4.15). Using this expression, the integral of i_{ri} at the kth cycle is written as below for this harvester:

$$\int i_{ri}(t)dt(k) = \left(C_{max} - C_{vi}(k)\right)(V_{ri} - V_F). \tag{4.49}$$

$C_{vi}(k)$ is expressed in (4.24). Replacing the value of $V_{rf}(k)$ as in (4.25) in (4.24), $C_{vi}(k)$ is rewritten as below:

$$C_{vi}(k+1) = \frac{V_{Cs}(k+1) + V_F}{V_{ri} - V_F}C_{min} \xrightarrow{k \to k-1} C_{vi}(k) = \frac{V_{Cs}(k) + V_F}{V_{ri} - V_F}C_{min}. \tag{4.50}$$

Replacing the above value of $C_{vi}(k)$ in (4.49):

$$\int i_{ri}(t)dt(k) = C_{max}(V_{ri} - V_F) - C_{min}\left(V_{Cs}(k) + V_F\right)$$

$$\xrightarrow{V_{Cs}(k) \ in \ (4.28)} \int i_{ri}(t)dt(k) = C_{min}(V_{ri} - V_F)(n - 1)r^k, \tag{4.51}$$

where r is defined in (4.32).

The current that goes through D_1 in Fig. 4.10 is the investment current. Unlike the reimbursement current that goes through the inductor every time that the flyback circuit activates, the investment current goes through D_1 in the investment phase of every operating cycle. Therefore, the total integral of the investment current before the flyback circuit activates is found as follows:

$$\int i_{ri}(t)dt = \sum_{k=k_1}^{k=k_2} \int i_{ri}(t)dt(k) = C_{min}(V_{ri} - V_F)(n-1) \sum_{k=k_1}^{k=k_2} r^k$$

$$\Rightarrow \int i_{ri}(t)dt = C_{min}(V_{ri} - V_F)(n-1)\frac{r^{k_1} - r^{k_2+1}}{1-r}. \tag{4.52}$$

Replacing the result of (4.35) in the above expression:

$$\int i_{ri}(t)dt = (C_S + C_{min})(V_2 - V_1). \tag{4.53}$$

The difference between the integral of the reimbursement and the investment currents is used in (2.28) to find E_{net} in C_r. Using (4.48) and (4.53), this parameter is found as below:

$$\int i_{rr}(t)dt - \int i_{ri}(t)dt = \frac{V_2 - V_1}{2(V_{ri} + V_F)}\left(C_S(V_2 + V_1 - 2V_{ri}) - 2C_{min}(V_{ri} + V_F)\right). \tag{4.54}$$

The flyback circuit activates every K_t cycles and transfers the accumulated energy in C_S back to C_r. Therefore, replacing the above value in (2.28), the average net generated energy in C_r in each cycle is found as follows:

$$E_{net} = \frac{1}{2K_t}\left(\int i_{rr}dt - \int i_{ri}dt\right)\left(2V_{ri} + \frac{1}{C_r}\left(\int i_{rr}dt - \int i_{ri}dt\right)\right) =$$

$$\frac{V_2 - V_1}{4K_t(V_{ri} + V_F)}\left(C_S(V_2 + V_1 - 2V_{ri}) - 2C_{min}(V_{ri} + V_F)\right)$$

$$\left(2V_{ri} + (V_2 - V_1)\frac{C_S(V_2 + V_1 - 2V_{ri}) - 2C_{min}(V_{ri} + V_F)}{2C_r(V_{ri} + V_F)}\right). \tag{4.55}$$

To maximize the storable energy in C_r at V_{ri}, C_r is selected to be much larger than C_S and C_{min}. Considering that $C_r \gg C_S, C_{min}$, the above expression may be simplified as follows:

$$E_{net} = \frac{V_{ri}(V_2 - V_1)}{2K_t(V_{ri} + V_F)}\left(C_S(V_2 + V_1 - 2V_{ri}) - 2C_{min}(V_{ri} + V_F)\right). \tag{4.56}$$

4.4   Optimizing the Net Generated Energy

The average net generated energy in the asynchronous harvester depends on C_S, the voltage limits, V_1, V_2, and K_t for specific values of C_{min}, V_{ri}, and V_F. This dependency is evident in (4.56). The voltage limits V_1 and V_2 may be adjusted by changing the frequency of the flyback circuit activation ($1/K_t$) and the duration that S_1 in Fig. 4.9 turns on. C_S defines the value of r in (4.32) and ultimately impacts how fast the voltage across C_S evolves. In this section, the values of C_S, V_1, V_2, and K_t for which the net generated energy is maximized are investigated. This optimization is done while all the conduction losses are considered.

4.4.1   Basic Steps

The voltage limits V_1 and V_2 are related to each other with the frequency of the flyback circuit activation ($1/K_t$) and the value of r. Following the below steps, make the optimization procedure possible.

The voltage limits may be expressed as follows:

$$V_1 = \big(1 + \alpha\,(n-1)\big)(V_{ri} - V_F) - V_F,$$
$$V_2 = \big(1 + \beta\,(n-1)\big)(V_{ri} - V_F) - V_F, \tag{4.57}$$

where α and β are constant coefficients and:

$$0 < \alpha < \beta < 1. \tag{4.58}$$

These voltage limits are mentioned similar to above in [2] without considering the forward voltage drop across the diodes. Using the above expressions, E_{net} in (4.56) may be written as follows:

$$E_{net} = \frac{V_{ri}(V_{ri} - V_F)(n-1)(\beta - \alpha)}{2K_t(V_{ri} + V_F)}\Big(C_S(V_{ri} - V_F)(n-1)(\beta + \alpha) -$$

$$4C_S V_F - 2C_{min}(V_{ri} + V_F)\Big). \tag{4.59}$$

Comparing the voltage limits in (4.57) with (4.34), the following result is obtained:

$$\begin{cases} n - (n-1)r^{k_1} = 1 + \alpha(n-1) \\ n - (n-1)r^{k_2+1} = 1 + \beta(n-1) \end{cases} \Rightarrow \frac{1-\beta}{1-\alpha} = \frac{r^{k_2+1}}{r^{k_1}} = r^{k_2+1-k_1} = r^{K_t},$$

$$\frac{1-\beta}{1-\alpha} = r^{K_t} \Rightarrow \quad \beta = 1 - r^{K_t} + \alpha r^{K_t}, \tag{4.60}$$

where r is defined in (4.32). Based on the above expression, K_t may be expressed as below:

$$K_t = \frac{\ln\left(\frac{1-\beta}{1-\alpha}\right)}{\ln(r)}. \tag{4.61}$$

Using the above expressions for the voltage limits and the relation between α, β, r, and K_t, the net generated energy in (4.56) is rewritten as follows:

$$E_{net} = \frac{V_{ri}(V_{ri} - V_F)(n-1)(1 - r^{K_t})(1 - \alpha)}{2K_t(V_{ri} + V_F)}.$$

$$\left(C_S(V_{ri} - V_F)(n-1)\left(1 - r^{K_t} + \alpha(1 + r^{K_t})\right) - \right.$$

$$\left. 4C_S V_F - 2C_{min}(V_{ri} + V_F)\right). \tag{4.62}$$

4.4.2 Optimization for a Specific Case

The basic steps in the previous subsection simplify the optimization significantly. Using the relation of (4.61) in (4.62), E_{net} depends on the values of α, r, and K_t for specific values of C_{min}, V_{ri}, and V_F. Therefore, the optimization is still a non-trivial task, since E_{net} depends on three parameters that are not directly relevant to each other. To tackle this issue, a medium step is considered in this section. This step is to optimize E_{net} for specific values of C_{min}, C_S, V_{ri}, V_F, and K_t firstly. With the specific values for these parameters, the net generated energy would only need to be optimized against changes in α, according to (4.62). Therefore, the optimal value of α may be found as follows:

$$\frac{dE_{net}}{d\alpha} = 0 \Rightarrow \alpha_{opt} = \frac{r^{K_t}}{1 + r^{K_t}} + \frac{2C_S V_F + C_{min}(V_{ri} + V_F)}{C_S(V_{ri} - V_F)(n-1)(1 + r^{K_t})}. \tag{4.63}$$

Using (4.60), the optimal value of β is obtained as follows:

$$\beta_{opt} = \frac{1}{1 + r^{K_t}} + \frac{2C_S V_F + C_{min}(V_{ri} + V_F)r^{K_t}}{C_S(V_{ri} - V_F)(n-1)(1 + r^{K_t})}. \tag{4.64}$$

4.4.3 Optimization

Based on the above result, the optimal values of α and β follow the below relation for specific values of C_{min}, C_S, V_{ri}, V_F, and K_t:

$$\alpha_{opt} + \beta_{opt} = 1 + \gamma, \qquad \gamma = \frac{2V_F + \frac{C_{min}}{C_S}(V_{ri} + V_F)}{(V_{ri} - V_F)(n-1)}. \tag{4.65}$$

Using the above relation between optimal values of α and β in (4.59), E_{net} may be expressed as follows:

$$E_{net} = \eta_1 \eta_2 \eta_3, \tag{4.66}$$

where:

$$\eta_1 = -C_S ln(r) \xrightarrow{C_S = \sigma C_{min}} \eta_1 = C_{min}\left(-\sigma . ln\left(\frac{\sigma}{\sigma+1}\right)\right),$$

$$\eta_2 = \frac{1 + \gamma - 2\beta}{ln\left(\frac{1-\beta}{\beta-\gamma}\right)},$$

$$\eta_3 = \frac{V_{ri}(V_{ri} - V_F)(n-1)}{2(V_{ri} + V_F)}\left((V_{ri} - V_F)(n-1) - 2V_F - \frac{C_{min}}{C_S}(V_{ri} + V_F)\right). \tag{4.67}$$

Figure 4.13a shows plot of η_1 against σ. This coefficient is bigger for higher values of σ. Based on the above formula, the value of C_S should be increased to have a higher value of η_1. Therefore, η_1 is bigger for higher values of C_S. Figure 4.13b shows the plot of η_2 against β for different values of γ. As can be seen, η_2 is maximized for lower values of γ. Based on the above expression, γ is less for higher values of C_S. Therefore, C_S should be chosen as high as possible to maximize η_2. Finally, η_3 is bigger for higher values of C_S, based on the above expression. All of the three coefficients in (4.66) increase for higher values of C_S. Therefore, the below procedure should be followed to optimize the voltage limits, C_S, and K_t for specific values of C_{min}, V_{ri}, and V_F.

- Select C_S as high as possible to maximize η_1, η_2, and η_3.
- Find the equivalent value of γ from (4.65) for the selected C_S.
- Find the optimal value of β for which η_2 is maximized in Fig. 4.13b for the relevant value of γ that is calculated in the previous step .
- Find the optimal value of α using the obtained optimal value of β and from (4.65).
- Find the value of K_t from (4.61) for the optimal values of α and β.

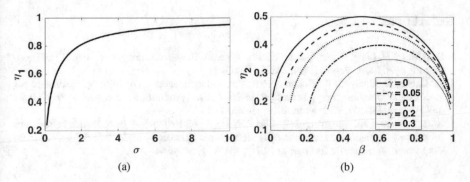

Fig. 4.13 Plot of (**a**) η_1 against σ and (**b**) η_2 against β for different values of γ

Figure 4.13b shows the plot of η_2 for few selected values of γ. This parameter may be plotted for any value of γ based on (4.67). This way, the optimal value of β may be found for different values of γ. Repeating the above steps, it is revealed that the optimal value of β is bigger than 0.5 and the optimal K_t is always 1 for $0 \leq \gamma < 1$. This implies that the flyback circuit should be activated in each operating cycle.

4.5 Conclusion

In [3], it is shown that the optimal net generated energy in the asynchronous harvester is half of the net generated energy in an equivalent synchronous harvester. The net generated energy in the equivalent synchronous harvester depends on the values of C_{min}, n, V_{ri}, and V_F. The gate pulse of the flyback switch is synchronized with the moment that the capacitance of C_V is minimal in this harvester. This adds complexity and power consumption to the circuit implementation. However, the control circuit can be implemented with a very low power. In the asynchronous circuit, the flyback circuit is not synchronized with the status of the variable capacitor and the implementation of the control circuit is simplified. The net generated energy depends on not only all parameters mentioned for the synchronous harvester but also on the capacitance of intermediate storage capacitor between the variable capacitor and the flyback circuit (C_S), the flyback circuit activation frequency, and the upper and lower voltage limits for the voltage across C_S. The optimal values for these parameters are explored in this section, while the conduction losses are considered.

References

1. B.C. Yen, J.H. Lang, A variable capacitance vibration-to-electric energy harvester. IEEE Trans. Circuits Syst. I Reg. Papers **53**(2), 288295 (2006)
2. A. Dudka, D. Galayko, E. Blokhina, P. Basset, Smart integrated conditioning electronics for electrostatic vibration energy harvesters, in *2014 IEEE International Symposium on Circuits and Systems (ISCAS)*, Melbourne, 2014, pp. 2600–2603
3. S.H. Daneshvar, M. Maymandi-Nejad, M.R. Yuce, J.-M. Redouté, A performance comparison between synchronous and asynchronous electrostatic harvesters, in *2019 IEEE International Conference on Industrial Technology (ICIT)*, 2019, pp. 349–354

Chapter 5
Electrostatic Harvesters with Inductor

Abstract Typical electrostatic harvesters that use an inductor in their investment and/or reimbursement circuitries are detailed in this chapter. These harvesters often require a large inductance value to operate optimally. The inductor is a bulky component and its presence is undesired in applications with small form-factors, e.g. medical applications. The ultimate goal of this chapter is to evaluate the performance of this type of electrostatic harvester when their inductor is miniaturized. Subsequently, an electrostatic harvester that operates optimally using a miniature inductor is explained. The steps for developing an experimental setup and practical design considerations for these harvesters are discussed in this chapter. Finally, the performance of the harvester that operates optimally with a miniaturized inductor is theoretically and experimentally compared with two conventional state-of-the-art electrostatic harvesters.

5.1 Preliminary Analysis Steps

The conduction losses are categorized to resistive and diode forward voltage related. These losses depend on the current that passes through the conducting path. The impact of series resistance of the components on this current becomes negligible when a large inductor exists in a conducting path. However, this impact is not negligible for smaller inductance values. Therefore, a detailed study of the current for smaller inductance values is an essential element in evaluating electrostatic harvesters with a miniaturized inductor. This section covers the fundamentals for this analysis.

5.1.1 Inductor-Based Charge Transfer

The general circuit structure of a sustainable electrostatic harvester is depicted in Fig. 2.2. The harvesters that are discussed in this chapter use an inductor in their

© The Author(s), under exclusive license to Springer Nature Switzerland AG 2022   157
S. H. Daneshvar et al., *Design of Miniaturized Variable-Capacitance Electrostatic Energy Harvesters*, https://doi.org/10.1007/978-3-030-90252-0_5

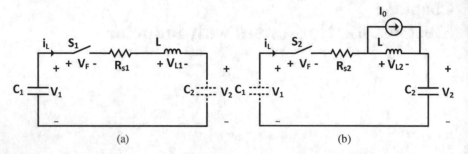

Fig. 5.1 General circuit topology in (**a**) energizing phase and (**b**) de-energizing phase

investment and/or reimbursement circuitries. The study here considers the impact of miniaturizing this inductor on the performance of these harvesters.

The energy is generated in an electrostatic harvester as the result of charges being transferred between C_r and C_V back and forth as explained in Sect. 2.2. Regardless of the exact circuit topology, the inductor-based charge transfer between capacitors occurs in two phases, i.e. energizing phase and de-energizing phase. Figure 5.1 shows representative components that exist in these phases. In this figure, R_{s1} and R_{s2} represent the series equivalent resistance of all the components in the energizing and the de-energizing phases. S_1 and S_2 represents any type of switch that initiates each of these phases, when they turn on.

During the energizing phase, an amount of charge is obtained from C_1. Therefore, C_1 should be present in the conducting path of the energizing phase. However, C_2 may or may not be present in this path, and therefore, it is depicted with dashed lines in Fig. 5.1a. During the de-energizing phase, an amount of charge is transferred to C_2. Therefore, C_2 should be present in the conducting path of the de-energizing phase. However, C_1 may or may not be present in this path, and therefore, it is depicted with dashed lines in Fig. 5.1b.

An inductor is present in the energizing path and its current increases from zero to a positive value, I_0. The same inductor should be in the de-energizing path and its current decreases from I_0 to zero. The direction of i_L in Fig. 5.1a and b is determined based on the assumption that at the end of consecutive energizing and de-energizing phases an amount of charge is transferred from C_1 to C_2. Based on this assumption, details of these phases are explained as follows for two cases, i.e. large inductance and small inductance.

5.1.2 Large Inductance

In applications where the form-factor is not a limiting parameter, the inductor is selected the largest possible to minimize the resistive conduction losses. In case that the inductance value is large enough to satisfy the following relations, these losses

Fig. 5.2 The current of a large inductor in consecutive energizing and de-energizing phases

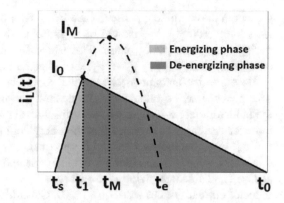

become negligible compared to the diode forward voltage-related conduction losses:

$$R_{s1}^2 \ll \frac{4L}{C_{t1}}, \quad R_{s2}^2 \ll \frac{4L}{C_{t2}}, \tag{5.1}$$

where C_{t1} and C_{t2} are the series equivalent of all the capacitances that are present in the energizing and the de-energizing phases, respectively.

Figure 5.2 shows the current of a large inductor during consecutive energizing and de-energizing phases. The energizing phase starts at t_s and the inductor current increases during the energizing phase. A combination of switches ends this phase at $t = t_1$ and the de-energizing phase starts during which the inductor current decreases. The amount of current that is stored in the inductor at the end of the energizing phase (I_0) is the initial current of the inductor at the beginning of the de-energizing phase, $i_L(t_1) = I_0$. The de-energizing phase ends at $t = t_0$, when the inductor current reaches zero, $i_L(t_0) = 0$. The inductor current is expressed as follows, if similar conditions as in (5.1) are true for the conducting path:

$$i_L(t) = \begin{cases} \dfrac{V_{L1}}{L\omega_1} sin(\omega_1(t - t_s)) & t_s < t < t_1 \\[4mm] I_0 cos(\omega_2(t - t_1)) - \dfrac{V_{L2}}{L\omega_2} sin(\omega_2(t - t_1)) & t_1 < t < t_0 \end{cases}, \tag{5.2}$$

where:

$$\omega_1 = \frac{1}{\sqrt{LC_{t1}}}, \quad \omega_2 = \frac{1}{LC_{t2}}, \tag{5.3}$$

In (5.2), V_{L1} and V_{L2} are the initial voltages across the inductor at the beginning of energizing and de-energizing phases, respectively. These voltages are specified in Fig. 5.1a and b. In these figures, V_F is the voltage drop across S_1 or S_2 when they turn on to initiate each of these phases. Since the inductor current is increasing during the energizing phase, V_{L1} should be positive. Therefore, C_2 should not be present in the energizing phase in case $V_1 \leq V_F + V_2$ is true at the beginning of this phase. Accordingly, V_{L2} should be negative as the inductor current is decreasing during the de-energizing phase. Therefore, C_1 should not be present in the de-energizing phase if $V_1 \geq V_F + V_2$, at the beginning of this phase.

Based on the amount of charge that should be transferred from C_1 to C_2, the time $t = t_1$ may be calculated. The dashed line in this figure shows the shape of the inductor current if there was no switching at $t = t_1$. Based on the shape of the inductor current, it is revealed that the energizing phase may not continue beyond $t = t_M$, when the inductor current is maximal, $i_L(t_M) = I_M$. This is due to the fact that after $t = t_M$, the inductor current starts decreasing. In the case of no switching at $t = t_1$, the de-energizing phase starts at $t = t_M$ and ends at $t = t_e$, when the inductor current reaches zero, $i_L(t_e) = 0$. Considering $i_L(t)$ in (5.2):

$$t_M = t_s + \frac{\pi}{2\omega_1}, \quad I_M = \frac{V_{L1}}{L\omega_1} \tag{5.4}$$

$$t_0 = t_1 + \frac{1}{\omega_2}\cos^{-1}\left(\frac{\frac{V_{L2}}{L\omega_2}}{\sqrt{I_0{}^2 + \left(\frac{V_{L2}}{L\omega_2}\right)^2}}\right). \tag{5.5}$$

5.1.3 Small Inductance

In applications with limiting form-factor, the inductor size is reduced. Consequently, the impact of the series resistance of the components on the conducting losses is not negligible. The current in the energizing and the de-energizing phases is calculated as follows when the conditions in (5.1) are not true.

Energizing Phase Figure 5.1a shows the representative components that exist in this phase. The stored energy in the inductor at the beginning of this phase is zero: The inductor initial current is zero, $i_L(t = t_s) = 0$. During this phase, i_L increases in the direction that is marked in this figure. It is useful to define the following parameter to study i_L:

$$k_1 = \frac{2}{R_{s1}}\sqrt{\frac{L}{C_{t1}}} = 2Q_{s1}, \tag{5.6}$$

where Q_{s1} is the series quality factor of R_{s1}, L, and C_{t1}.

The current that goes through the inductor during this phase is

$$i_L(t) = \begin{cases} \dfrac{V_{L1}}{L\omega_1} sin(\omega_1(t-t_s))e^{-\frac{R_{s1}}{2L}(t-t_s)} & k_1 > 1 \\[4mm] \dfrac{V_{L1}}{2L\omega_1}\left(e^{\left(-\frac{R_{s1}}{2L}+\omega_1\right)(t-t_s)} - e^{\left(-\frac{R_{s1}}{2L}-\omega_1\right)(t-t_s)}\right) & k_1 < 1 \end{cases} , \quad t_s < t < t_1,$$

(5.7)

where:

$$\omega_1 = \frac{\sqrt{\left|\frac{4L}{C_{t1}} - R_{s1}^{2}\right|}}{2L}.$$

(5.8)

To find the moment when the inductor current reaches its maximal value ($t = t_M$), the equation $i_L'(t) = 0$ should be solved for each of the above cases: $k_1 > 1$ and $k_1 < 1$. The result is

$$t_M = \begin{cases} t_s + \dfrac{1}{\omega_1}cos^{-1}\left(\dfrac{\frac{R_{s1}}{2L}}{\sqrt{\omega_1^2 + \frac{R_{s1}^2}{4L^2}}}\right) & k_1 > 1 \\[6mm] t_s + \dfrac{1}{2\omega_1}\ln\left(\dfrac{-\frac{R_{s1}}{2L} - \omega_1}{-\frac{R_{s1}}{2L} + \omega_1}\right) & k_1 < 1 \end{cases}.$$

(5.9)

Figure 5.3a shows the inductor current during possible energizing phase ($t_s < t < t_M$) with solid lines and during dc-energizing phase with dashed lines in case no switching occurs at $t_s < t_1 < t_M$. This current is shown for different values of k_1 to illustrate the impact of this parameter on the shape of the inductor current. It can be seen that the maximum inductor current (I_M) is less for higher values of k_1. As expected from (5.7), the inductor current follows the sinusoidal shape for higher values of k_1 and exponential shape for lower values of k_1.

Figure 5.3b shows the integral of $i_L(t)$ during the possible energizing phase ($t_s < t < t_M$) for different values of k_1. The integral of $i_L(t)$ from $t = t_s$ to $t = t_1$ is equal to the amount of charge that is obtained from C_1 during this phase. Based on this figure, it is possible to obtain more charge from C_1 for higher values of k_1. Figure 5.3 is depicted for $V_L = 1$ V, $C_{t1} = 1$ nF, $R_{s1} = 10$ Ω, and the inductance can be calculated based on (5.6): L varies between 1 nH and 56.25 nH.

De-energizing Phase Figure 5.1b shows the representative components that exist in this phase. An amount of energy is stored in the inductor during the energizing phase, hence the inductor current is non-zero at the beginning of the de-energizing phase, $i_L(t = t_1) = I_0$. During this phase, the stored current in the inductor flows into the capacitor that receives charges (C_2), until it reaches zero: $i_L = 0$. The current that goes through the inductor during this phase is

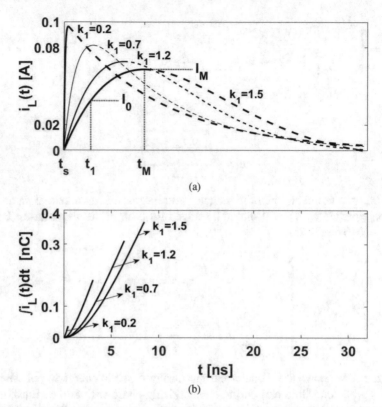

Fig. 5.3 (**a**) The energizing current for different values of k (**b**) The integral of the energizing current (charge) for different values of k

$$i_L(t) = \begin{cases} e^{-\frac{R_{s2}}{2L}(t-t_1)}\left(a_1 cos(\omega_2(t-t_1)) - a_2 sin(\omega_2(t-t_1))\right) & k_2 > 1 \\ a_1 e^{(-\frac{R_{s2}}{2L}+\omega_2)(t-t_1)} + a_2 e^{(-\frac{R_{s2}}{2L}-\omega_2)(t-t_1)} & k_2 < 1 \end{cases} , t_1 < t < t_0,$$

(5.10)

where k_2 and ω_2 are obtained from (5.6) and (5.8), respectively, where R_{s1} and C_{t1} are replaced by R_{s2} and C_{t2} and:

$$\begin{cases} a_1 = I_0, \quad a_2 = \frac{R_{s2}}{2L\omega_2}I_0 + \frac{V_{L2}}{L\omega_2} & k_2 > 1 \\ a_1 = I_0\left(\frac{1}{2} - \frac{R_{s2}}{4L\omega_2}\right) - \frac{V_{L2}}{2L\omega_2}, \quad a_2 = I_0\left(\frac{1}{2} + \frac{R_{s2}}{4L\omega_2}\right) + \frac{V_{L2}}{2L\omega_2} & k_2 < 1 \end{cases}$$

(5.11)

Figure 5.4a shows the inductor current during the de-energizing phase for different values of k_2. The initial inductor current for all of these plots is the same, I_0. As expected from (5.10), the inductor current follows an exponential shape for lower values of k_2 and follows a sinusoidal shape for higher values of

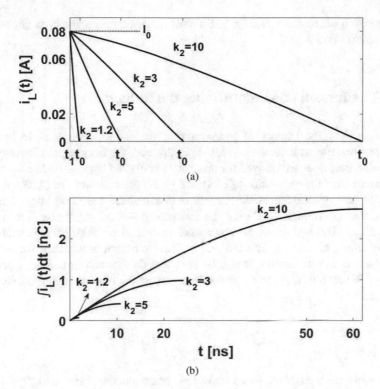

Fig. 5.4 (a) The de-energizing current for different values of k (b) The integral of the de-energizing current (charge) for different values of k

k_2. Figure 5.4b shows the integral of the inductor current during the de-energizing phase for different values of k_2. The integral of the inductor current is equal to the amount of charge that is transferred to C_2 during this phase. As can be seen, it is possible to transfer more charge to C_2 for higher values of k_2.

The de-energizing phase ends at $t = t_0$, when the inductor current reaches zero: $i_L(t = t_0) = 0$. These moments are marked in Fig. 5.4a for different values of k_2. By equating $i_L(t)$ (as expressed in (5.10)) to zero, this parameter is found:

$$
t_0 = \begin{cases} t_1 + \dfrac{1}{\omega_2}\cos^{-1}\left(\dfrac{a_2}{\sqrt{a_1{}^2 + a_2{}^2}}\right) & k_2 > 1 \\[4mm] t_1 + \dfrac{1}{2\omega_2}\ln\left(\dfrac{-a_2}{a_1}\right) & k_2 < 1 \end{cases} \tag{5.12}
$$

where a_1 and a_2 are found in (5.11).

The component values for plotting Fig. 5.4 are $C_{t2} = 1\,\text{nF}$, $R_{s2} = 10\,\Omega$, and $I_0 = 80\,\text{mA}$. To illustrate the impact of k_2 on the shape of the inductor current in de-energizing phase more clearly, a different range of values for k_2 is considered compared to Fig. 5.3.

5.2   The Impact of Miniaturizing the Inductor

In this section, the impact of miniaturizing the inductor on the electrostatic harvesters that use an inductor is studied. To this end, the net generated energy and the conduction losses are derived for three state-of-the-art harvesters. In calculating these parameters the expressions (2.28) and (2.29) are followed. In (2.28), the net generated energy in electrostatic harvesters is calculated based on integral of the currents that go through C_r during the investment and the reimbursement phases (i_{ri} and i_{rr}). These currents are expressed in Sect. 5.1.1 for different values of the coefficient, k_1 and k_2. The integral of these currents is obtained for each of the harvesters in this section. It will be seen that these integrals can be expressed based on the ratio of C_{min} to C_r, the voltage across the inductor and the following coefficient:

$$k_m = \frac{2}{R_s}\sqrt{\frac{L}{C_{min}}} = 2Q_{sm}, \tag{5.13}$$

where R_s is the equivalent series resistance of conducting paths and Q_{sm} is the series quality factor of L, R_s, and C_{min}. This coefficient is defined similar to k_1 in (5.6) with the minimal capacitance of C_V as the capacitance of the path.

The efficiencies of the variable-capacitance harvesters that use an inductor depend on the resistance in the conducting paths (R_s). Therefore, transistors in this type of harvester should be selected not only to tolerate the voltages and currents but also to keep R_s as minimal as possible. Selecting the transistors, the constant capacitors and the variable capacitor, C_{min} and R_s are determined in the ratio of k_m in (5.13). Therefore, k_m varies according to employing different inductor values, L. Studying the impact of scaling down k_m generalizes the presented analysis for different quality factors of C_{min}, R_s, and L in series instead of individual values of these components. This implies that increasing R_s has the same impact as decreasing L on the net generated energy in this type of harvester.

5.2.1   An Inductor-Based V:C-A:S Harvester

An electrostatic harvester that operates under the voltage-constraint scheme during the recovery phase and under the charge-constraint scheme during the harvesting phase is examined in this section. This harvester has an asynchronous switching

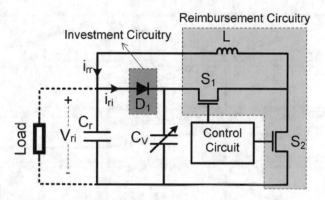

Fig. 5.5 The V:C-A:S electrostatic harvester presented in [1]

event at the start of the investment phase and a synchronous switching event at the start of the reimbursement phase.

5.2.1.1   The Circuit, Signals, and QV Diagram

Figure 5.5 shows the harvester presented in [1]. In this circuit, C_r is the storage component and is initially charged to V_{ri}. The investment and the reimbursement circuitries are highlighted in Fig. 5.5 for this harvester. As can be seen, the investment circuitry is a single diode and does not need any gate pulses from the control circuit. Therefore, the switching event to start the consecutive investment and harvesting phases is asynchronous for this harvester. However, the control circuit should generate gate pulses for the transistors in the reimbursement circuitry. These gate pulses should be synchronized with the moment that $C_V = C_{min}$, and therefore, the switching event to start the consecutive reimbursement and recovery phases is synchronous. The phases of operation, the voltage across the variable capacitor, the voltage across C_r, and the current in C_r during these phases are shown in Fig. 5.6. The QV diagram of this harvester is shown in Fig. 5.7. Following these figures the operation of the harvester is explained here below.

Harvesting Phase: The variable capacitor (C_V) is charged to V_r at the beginning of this phase and is isolated from the rest of the circuit during this phase: D_1 and S_1 are off. Hence, the charge in this capacitor is kept constant. The energy source then changes the capacitance of the variable capacitor from C_{max} to C_{min}. Based on $Q = CV$, the voltage across the variable capacitor increases to nV_r:

$$V_{C_V} : V_r \rightarrow nV_r, \quad n = \frac{C_{max}}{C_{min}}. \tag{5.14}$$

The start and the end of the harvesting phase are marked by **b** and **c**, respectively, in Fig. 5.7.

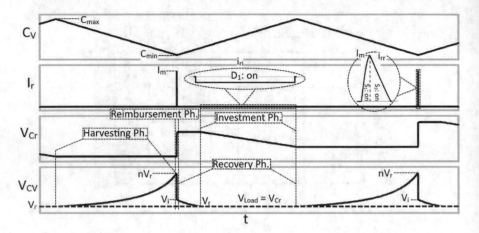

Fig. 5.6 Phases of operation, the voltages across capacitors, and the current passing through C_r in the V:C-A:S harvester

Reimbursement Phase: The reservoir capacitor (C_r) provides the initial charge to C_V and powers the load. This capacitor eventually depletes in case that it is not recharged. To make the system sustainable, the harvested energy in the variable capacitor should be transferred back to C_r. This charge transfer occurs in the consecutive energizing and de-energizing phases as explained in Sect. 5.1.1. The energizing phase starts when S_1 turns on and C_V connects to C_r through an inductor. Charge is transferred from C_V to C_r until the current in the inductor reaches its maximum value and S_1 turns off. At this moment, the de-energizing phase starts and the stored energy in the inductor transfers to C_r through S_2 until the inductor current reaches zero. The reimbursement current (i_{rr}) that goes through C_r during this phase is magnified in Fig. 5.6. In this figure, the switch that is on in each period of time is specified; other switches are off during that time. During this phase, the voltage across the variable capacitor decreases from nV_r to V_r, when a large inductor is used. However, this voltage decreases from nV_r to V_i (a voltage higher than V_r) when a miniature inductor is used. This consideration is applied in the derived formulas to evaluate the impact of scaling the inductor size on the operation of this harvester.

In Fig. 5.7, point **c** and **d** shows the start and the end of the reimbursement phase, in case a miniature inductor is used. In case that a large inductor is used, the reimbursement phase starts at point **c** and ends at point **a'**.

Recovery Phase: The energy source starts changing the capacitance of the variable capacitor from C_{min} towards C_{max}, while C_V is isolated from the rest of the circuit. Accordingly, the voltage across the variable capacitor decreases from V_i towards V_r. This phase ends when C_V reaches its maximum value, i.e. C_{max}. The variable capacitor becomes ready for the next operating cycle at the end of this phase.

Fig. 5.7 QV diagram of the V:C-A:S harvester that is depicted in Fig. 5.5

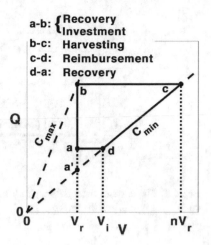

Investment Phase: During this phase, C_r charges C_V to an initial voltage to make the variable capacitor ready for the harvesting phase. The investment phase starts when the voltage across C_V reaches V_r during the recovery phase, as shown in Fig. 5.6. At this moment, D_1 turns on and C_r starts charging C_V: The investment current (i_{ri}) that goes through C_r is magnified in Fig. 5.6. In this figure, the voltage across C_r is zoomed in to illustrate its variations.

The recovery phase starts at point **d** on Fig. 5.7, when a miniature inductor is used. This phase starts at point **a'**, if a large inductor is used. At point **a** (or **a'**), when the voltage across C_V is equal to V_r, the investment phase starts. Both the recovery and the investment phases end at point **b**, when the capacitance of the variable capacitor is maximal.

As can be seen, this harvester operates under the charge-constraint scheme during the harvesting phase. The harvester operates under voltage-constraint scheme during the recovery phase when a large inductor is used. Therefore, this harvester is categorized as V:C. However, the impact of employing a miniature inductor on this is that it operates under a combination of charge-constraint and voltage-constraint schemes during the recovery phase.

5.2.1.2   The Net Generated Energy

Figure 5.8 shows the currents that go through C_r during the investment and the reimbursement phases in this harvester. These currents were magnified on Fig. 5.6. The dots on Fig. 5.8 show the time lapse between i_{ri} and i_{rr} that is evident in Fig. 5.6.

The investment phase starts at the moment that the voltage across C_V becomes equal to V_r during the recovery phase. At this moment, the capacitance of the variable capacitor is C_i. This moment is marked as t_i in Fig. 5.8. The investment

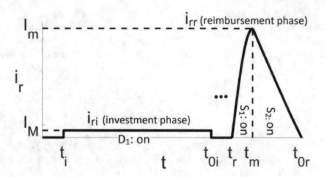

Fig. 5.8 The currents that go through C_r in the investment and the reimbursement phases in the circuit that is depicted in Fig. 5.5

phase ends at the moment that the capacitance of C_V reaches its maximal value. This moment is marked as t_{0i} in Fig. 5.8. During this phase, C_r is connected to C_V and since $C_r \gg C_{max}$, the voltage across C_V is kept constant. The voltage across C_V is kept constant at V_r if:

$$V_{ri} = V_r + V_F, \tag{5.15}$$

where V_{ri} is the initial voltage across C_r and V_F is the forward voltage drop across D_1. The integral of the investment current that goes through C_r, D_1, and C_V is obtained as follows:

$$\begin{cases} i_{ri} = V_r \dfrac{dC_V}{dt} \Rightarrow \displaystyle\int_{t_i}^{t_{0i}} i_{ri}dt = V_r \Delta C_V = V_r(C_{max} - C_i) \\ C_i = \dfrac{V_i}{V_r} C_{min}, \quad V_i > V_r \end{cases} \Rightarrow \int_{t_i}^{t_{0i}} i_{ri}dt = C_{min}V_r(n - \dfrac{V_i}{V_r}). \tag{5.16}$$

The reimbursement phase starts at the moment that the capacitance of C_V is minimal and the voltage across it is nV_r. This moment is marked as t_r in Fig. 5.8. During the energizing phase, S_1 is on and C_V is connected to C_r. This phase ends at the moment that the inductor current reaches its maximal value. This moment is marked as t_m in Fig. 5.8. As can be seen in Fig. 5.6, the voltage across C_V reaches V_i from nV_r during this phase, hence:

$$V_i = nV_r - \frac{1}{C_{min}} \int_{t_r}^{t_m} i_{rr}dt \Rightarrow \int_{t_r}^{t_m} i_{rr}dt = C_{min}(nV_r - V_i). \tag{5.17}$$

During the de-energizing phase, the remained current in the inductor goes through S_2 and C_r. This phase ends when the inductor current reaches zero. This moment is marked as t_{0r} in Fig. 5.8. From (5.16) and (5.17), it is revealed that:

$$\int_{t_i}^{t_{0i}} i_{ri} dt = \int_{t_r}^{t_m} i_{rr} dt \Rightarrow \int i_{rr} dt - \int i_{ri} dt = \int_{t_m}^{t_{0r}} i_{rr} dt. \tag{5.18}$$

Considering the above expression, the net generated in this harvester according to (2.28) is

$$E_{net} = \frac{1}{2} \int_{t_m}^{t_{0r}} i_{rr} dt \left(2(V_r + V_F) + \frac{1}{C_r} \int_{t_m}^{t_{0r}} i_{rr} dt \right). \tag{5.19}$$

As can be seen in the above expression, the integral of i_{rr} in $t_m \le t \le t_{0r}$ should be calculated to find E_{net} for this harvester.

The expressions for i_{rr} in this harvester are the same as in (5.10). In these expressions, I_0 should be replaced with I_M, as S_1 turns off at $t = t_M$ in the energizing phase. To find I_M, the relation $t = t_M$ is used in (5.7), where t_M is found in (5.9). Using the coefficient that is defined in (5.6) for this conducting path, I_M is expressed as follows:

$$I_M = \frac{2V_{L1}}{R_{s1}k_t} f(k_t), \quad f(k_t) = \begin{cases} e^{\frac{-1}{\sqrt{k_t^2-1}} \cos^{-1}\left(\frac{1}{k_t}\right)} & k_t > 1 \\ \left(\frac{1 + \sqrt{1 - k_t^2}}{1 - \sqrt{1 - k_t^2}}\right)^{\frac{-1}{2\sqrt{1-k_t^2}}} & k_t < 1 \end{cases}, \tag{5.20}$$

where R_{s1} is the equivalent series resistance of the energizing path and:

$$\begin{cases} V_{L1} = nV_r - V_{ri} = nV_r - V_r - V_F - V_S \\ \sigma_F = \dfrac{V_F}{V_r} \\ \sigma_S = \dfrac{V_S}{V_r} \end{cases}, \quad \Rightarrow V_{L1} = (n - 1 - \sigma_F - \sigma_S)V_r.$$

$$\tag{5.21}$$

where V_S is the overdrive voltage of transistor, S_1. Transistors should be biased in deep-triode region, while conducting, to decrease their conduction losses. The transistor may be modelled with a single on-resistance in this region, when it is conducting. The control circuit should generate an appropriate Gate–Source signal for this purpose and this is explained in Sect. 3.1.1. If this condition is met, the overdrive voltage of the transistor and subsequently σ_S would be zero in the above expression. The general case, where σ_S is not zero is considered for the transistors in this chapter. However, this parameter could be zero with the considerations in Sect. 3.1.1. In the above expressions, k_t may be replaced with k_m (defined in (5.13)) according to the following relation:

$$\begin{cases} k_t = \dfrac{2}{R_{s1}}\sqrt{\dfrac{L}{C_t}}, & R_{s1} = R_s \\[2mm] C_t = C_r || C_{min} = \dfrac{C_r C_{min}}{C_r + C_{min}} & \Rightarrow k_t = k_m\sqrt{1+\gamma}, \\[2mm] \gamma = \dfrac{C_{min}}{C_r} \end{cases} \qquad (5.22)$$

Knowing I_M and using (5.10), the integral of the i_{rr} in $t_m \leq t \leq t_{0r}$ is obtained as follows:

$$\int_{t_m}^{t_{0r}} i_{rr}dt =$$

$$\begin{cases} \dfrac{C_r V_r}{2}\left(\sqrt{k_r^2-1}\; e^{\frac{-1}{\sqrt{k_r^2-1}}cos^{-1}\left(\frac{a_2}{\sqrt{a_1^2+a_2^2}}\right)}\sqrt{a_1^2+a_2^2}+ \right. \\[4mm] \qquad\qquad\qquad\qquad\qquad\qquad \left. a_1 - a_2\sqrt{k_r^2-1}\right) \quad k_r > 1, \\[4mm] C_r V_r\left(\sqrt{1-k_r^2}\sqrt{-a_1 a_2}\left(\dfrac{-a_2}{a_1}\right)^{\frac{-1}{2\sqrt{1-k_r^2}}} + \dfrac{V_{L2}}{V_r}\right) \quad k_r < 1 \end{cases}$$
$$(5.23)$$

where:

$$\begin{cases} k_r = \dfrac{2}{R_{s2}}\sqrt{\dfrac{L}{C_r}}, & \Rightarrow k_r \approx k_m\sqrt{\gamma}, \\[2mm] R_{s1} = R_s \approx R_{s2} \end{cases} \qquad (5.24)$$

In the above expression, the equivalent series resistance of the energizing and de-energizing paths is considered to be approximately the same. In (5.23):

$$\begin{cases} a_1 = \dfrac{2V_{L1}}{k_t V_r} f(k_t) \\[3mm] a_2 = \dfrac{2}{\sqrt{k_r^2-1}}\left(\dfrac{V_{L1}}{k_t V_r} f(k_t) - \dfrac{V_{L2}}{V_r}\right) \end{cases}, \quad k_r > 1, \qquad (5.25)$$

where the same overdrive voltage, V_S, is considered for transistor, S_2.

$$\begin{cases} a_1 = \dfrac{V_{L1}}{k_t V_r} f(k_t) \left(1 - \dfrac{1}{\sqrt{1 - k_r^2}}\right) + \dfrac{V_{L2}}{V_r \sqrt{1 - k_r^2}} \\[4mm] a_2 = \dfrac{V_{L1}}{k_t V_r} f(k_t) \left(1 + \dfrac{1}{\sqrt{1 - k_r^2}}\right) - \dfrac{V_{L2}}{V_r \sqrt{1 - k_r^2}} \end{cases} , \quad k_r < 1, \qquad (5.26)$$

where $f(k_t)$ depends on the energizing phase and is defined in (5.20).

In all the above expressions, for this harvester:

$$V_{L1} = (n - 1 - \sigma_F - \sigma_S)V_r,$$
$$V_{L2} = -(1 + \sigma_F + \sigma_S)V_r. \qquad (5.27)$$

According to the above expressions, the integral of i_{rr} may be expressed as follows:

$$\int_{t_m}^{t_{0r}} i_{rr} dt = C_r V_r y_{r2}(k_m, n, \gamma, \sigma_F, \sigma_S), \qquad (5.28)$$

where y_{r2} is a function of the coefficients, k_m, n, γ, σ_F, and σ_S, and can be found by comparing the above expression with (5.23). Using the expression for E_{net} in (5.19) and the above expression:

$$E_{net} = \frac{1}{2} C_r V_r^2 y_{r2}(k_m, n, \gamma, \sigma_F, \sigma_S) \Big(2(1 + \sigma_F) + y_{r2}(k_m, n, \gamma, \sigma_F, \sigma_S)\Big). \qquad (5.29)$$

5.2.1.3 The Conduction Losses

In the steps of finding E_{net}, the value of V_i was not needed to be calculated explicitly. However, in finding the total conduction losses ($E_{loss-tot}$), this parameter should be calculated. Using (5.17) to find V_i, the integral of i_{rr} in $t_r \leq t \leq t_m$ should be calculated. The expressions for i_{rr} during this time are the same as expressed in (5.7). The energizing phase for this harvester continues until the moment that the inductor current reaches its maximal value, $t_1 = t_M$. The expressions for t_M are obtained in (5.9). Therefore, using (5.7) and (5.9), the integral of i_{rr} for the energizing phase is expressed:

$$\int_{t_r}^{t_m} i_{rr} dt =$$

$$\begin{cases} C_t V_r (n-1-\sigma_F - \sigma_S)\left(1 - \dfrac{2}{k_t} e^{\frac{-1}{\sqrt{k_t^2 -1}}\cos^{-1}\frac{1}{k_t}}\right) & k_t > 1 \\[4mm] C_t V_r (n-1-\sigma_F - \sigma_S)\dfrac{k_t^2}{2\sqrt{1-k_t^2}}\left(\left(\dfrac{1+\sqrt{1-k_t^2}}{1-\sqrt{1-k_t^2}}\right)^{\frac{1}{2}-\frac{1}{2\sqrt{1-k_t^2}}} \right. \\[4mm] \left. \left(\dfrac{4\left(1-\sqrt{1-k_t^2}\right)}{-\sqrt{1-k_t^2}\left(1+\sqrt{1-k_t^2}\right)}\right) + \dfrac{2\sqrt{1-k_t^2}}{k_t^2}\right) & k_t < 1 \end{cases} \quad , \qquad (5.30)$$

where k_t and C_t are defined in (5.22). The integral in the above expression may be expressed as follows:

$$\int_{t_r}^{t_m} i_{rr} dt = C_t V_r y_{r1}(k_m, n, \gamma, \sigma_F, \sigma_S). \qquad (5.31)$$

Using the above expressions and (5.16):

$$V_i = y_\gamma (k_m, n, \gamma, \sigma_F, \sigma_S) V_r, \qquad (5.32)$$

where:

$$y_\gamma = n - \frac{1}{1+\gamma} y_{r1}(k_m, n, \gamma, \sigma_F, \sigma_S). \qquad (5.33)$$

The deliverable energy (E_{del}) is found by calculating the enclosed area in Fig. 5.7 as follows:

$$E_{del} = \frac{1}{2} C_{max}(n-2)V_r^2 + \frac{1}{2} C_{min} V_i (2V_r - V_i), \qquad (5.34)$$

where V_i is expressed in (5.32) and E_{del} is re-written as follows:

$$E_{del} = \frac{1}{2} C_{min} V_r^2 y_\gamma (k_m, n, \gamma, \sigma_F, \sigma_S)\Big(2(n-1) + y_\gamma (k_m, n, \gamma, \sigma_F, \sigma_S)\Big). \qquad (5.35)$$

Using (2.29), the total conduction losses in this harvester are obtained where E_{del} is expressed in (5.34) and E_{net} is expressed in (5.29). Both E_{net} and E_{del} are expressed as functions of k_m, n, γ, σ_F, and σ_S in these expressions. Therefore, $E_{loss-tot}$ is also a function of these parameters.

5.2.1.4   Energy Plots

Figure 5.9 shows the deliverable energy (E_{del}), the net generated energy (E_{net}) and the total conduction losses ($E_{loss-tot}$) in this harvester, where $C_{min} = 1$ nF, $n = 5$,

Fig. 5.9 E_{del}, E_{net}, and $E_{loss-tot}$ for the V:C-A:S harvester

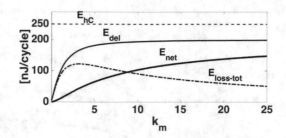

$C_r \gg C_{max}$, and $V_r = 5\ V \gg V_F$, V_S. These energies are plotted versus k_m to illustrate a generalized analysis for the impact of scaling the inductor as explained in Sect. 5.2. E_{hC} on this figure shows the amount of harvested energy. As can be seen, the deliverable energy is always less than this amount. E_{del} saturates to its maximal value for large inductance values (less than E_{hC}). The maximal deliverable energy for large inductance values is calculated from the enclosed area in Fig. 5.7, where $V_i = V_r$ as follows:

$$E_{del-max} = \frac{1}{2}C_{min}(n-1)^2 V_r^2. \tag{5.36}$$

The total conduction losses peak at lower values of k_m but eventually decreases for higher values of k_m (larger inductors). In this plot, the net generated energy (E_{net}) equals to the deliverable energy (E_{del}) for high inductance values as $\sigma_F \approx \sigma_S \approx 0$.

Figure 5.9 is plotted for the case that C_r is selected much larger than C_V, ($C_r \gg C_{max}$), hence γ equals zero in all the expressions for E_{net}, E_{del}, and $E_{loss-tot}$. In these expressions, σ_F and σ_S equal zero, since the forward voltage drop across D_1 and the overdrive voltage of S_1 and S_2 are negligible compared to V_r. Therefore, E_{net}, E_{del}, and $E_{loss-tot}$ depend on C_r, C_{min}, V_r, n, and k_m for this figure. Figure 5.10 shows the dependency of E_{net} on different values of n, where $C_r \gg C_{max}$, and $V_r = 5\ V \gg V_F$, V_S. In any technology that is used to fabricate a variable capacitor, the maximal feasible capacitance in a given volume has a practical limit. Therefore, C_{max} has a fixed value (5 nF) for all the plots in this figure.

In miniaturized harvesters, it is important NOT to select a large inductor as inductors are bulky. The impact of scaling the inductor on the net generated energy in the V:S-A:S harvester for different values of n is plotted in Fig. 5.10. As can be seen, the net generated energy does not increase much after a point in each plot. The value of k_m for which the net generated energy in the harvester is 90% of E_{del} is specified on each plot in this figure. A designer may select these points as an optimal inductance value. This is an important result of the presented analysis for studying the impact of scaling the inductance value on the net generated energy in electrostatic harvesters. Figure 5.10 is plotted as an example to illustrate the impact of scaling the inductor. The comprehensive expressions for E_{net} in this section can be used for different component values.

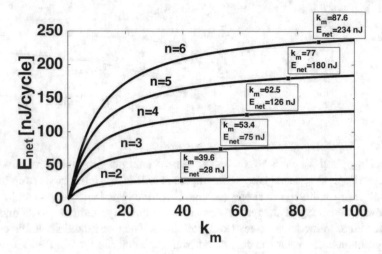

Fig. 5.10 The net generated energy in the V:C-A:S harvester versus k_m for different values of n

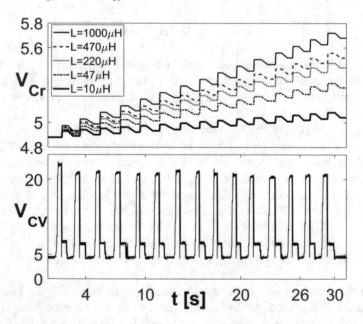

Fig. 5.11 The voltages across the reservoir capacitor and the variable capacitor in the V:C-A:S harvester for different inductance values based on the experimental results

5.2.1.5 Experimental Results

Details of an experimental setup for evaluating electrostatic harvesters are explained in Sect. 5.4. The V:C-A:S harvester in this section is tested using this setup. The results illustrate the impact of scaling the inductor on the performance of this harvester in practice. The theoretical results are then compared with these experimental results in Sect. 5.3.6.2.

Figure 5.11 shows the experimental results for this harvester when different inductors are used in the range of 10–1000 µH. In this figure, V_{Cr} is plotted based on the test results for different inductors and V_{CV} is the voltage across the variable capacitor in one of these experiments. As can be seen, more energy is stored in C_r when larger inductors are used. The variable capacitor is charged to 4.4 V at the beginning of each cycle, and this is achieved by adjusting the initial voltage across C_r.

5.2.2   An Inductor-Based C:C-S:S Harvester

An electrostatic harvester that operates under the charge-constraint scheme during both recovery and harvesting phases is examined in this section. This harvester has synchronous switching events at the start of the investment and the reimbursement phases.

5.2.2.1   The Circuit, Signals, QV Diagram

Figure 5.12 shows the charge-constraint variable-capacitance harvester presented in [2]. In this circuit, C_r is the storage component that is initially charged to V_{ri}. The harvester shares a single circuitry for implementing both investment and reimbursement phases. This circuitry is highlighted in Fig. 5.12. The control circuit generates the gate pulses for S_1 and S_2. These gate pulses should be synchronized with the moments that the capacitance of the variable capacitor is $C_V = C_{max}$ and $C_V = C_{min}$. Therefore, both the switching events to start the consecutive investment and harvesting phases and to start the consecutive reimbursement and recovery phases are synchronous. The phases of operation, the voltage across the variable capacitor, the voltage across C_r, and the current in C_r during these phases for this harvester are shown in Fig. 5.13. The QV diagram of this harvester is shown in Fig. 5.14. Following these figures, the operation of the harvester is explained here below.

Harvesting Phase: Point **b** on Fig. 5.14 shows the start of this phase. The variable capacitor is charged to V_r and its capacitance is maximal at the beginning of this phase. The transistors, S_1 and S_2 are off and therefore, C_V is isolated from the rest of the circuit during this phase. The voltage across C_V reaches nV_r at the end of this phase, the same way as explained for the V:C-A:S harvester in Sect. 5.2.1.1. Point **c** marks the end of this phase in Fig. 5.14. V_{CV} on Fig. 5.13 depicts the same changes in the voltage across C_V.

Reimbursement Phase: When the capacitance of the variable capacitor is minimal and the voltage across it is equal to nV_r, S_2 turns on and the current of the inductor

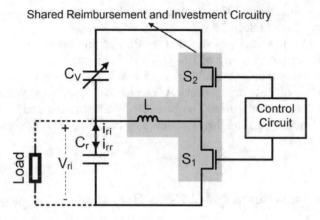

Fig. 5.12 The C:C-S:S electrostatic harvester presented in [2]

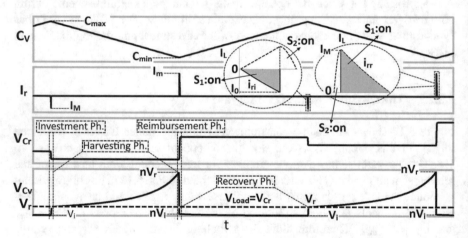

Fig. 5.13 Phases of operation, the voltages across capacitors, and the current passing through C_r in the C:C-S:S harvester

starts increasing. This switch turns off when the current of the inductor reaches its maximum value and S_1 turns on. The stored current in the inductor then goes through C_r until it reaches zero. The current that passes through L during this phase is magnified in Fig. 5.13, and the portion of it that goes through C_r is greyed and marked with i_{rr}. In each period of time in this figure, only the specified switch is on and the other one is off. During this phase, the voltage across C_V decreases from nV_r to 0, when a large inductor is used. However, this voltage decreases from nV_r to nV_i (a voltage higher than 0) when a miniature inductor is employed. This is considered in deriving formulas for evaluating the impact of scaling the inductor on the performance of this harvester. In Fig. 5.14, point **c** shows the start of this phase and point **a'** shows the end of this phase if a large inductor is used. In case that a miniature inductor is employed, point **d** shows the end of this phase.

Fig. 5.14 QV diagram of the
C:C-S:S harvester that is
depicted in Fig. 5.12

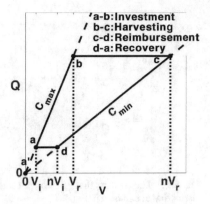

Recovery Phase: During this phase, the variable capacitor is isolated: S_1 and S_2 are off. Unlike the harvester in Fig. 5.5, there is no overlap between the recovery phase and the investment phase of this harvester, i.e. the investment phase starts when the recovery phase ends. Therefore, the voltage across the variable capacitor changes from nV_i to V_i during this phase. Point **d** shows the start of this phase and point **a** marks the end of this phase in case a miniature inductor is used. In case of using a large inductor, point **a'** shows the start of this phase. Since the voltage across C_V is zero at this point, the same point **a'** shows the end of this phase.

Investment Phase: When the capacitance of the variable capacitor is maximal, S_1 turns on until the inductor current reaches I_0 from zero. At this moment, S_2 turns on to pass the stored current in the inductor through C_V and turns off when $i_L = 0$. The inductor current (I_L) is magnified in Fig. 5.13 for this phase and the portion of it that goes through C_r is greyed and marked with i_{ri}. In Fig. 5.14, point **b** shows the end of this phase. Either points **a** or **a'** marks the start of this phase depending on whether or not a miniature inductor is used.

5.2.2.2   The Net Generated Energy

According to (2.28), the integral of the currents i_{ri} and i_{rr} should be calculated to find E_{net} for this harvester. Figure 5.15 shows the inductor current during the investment and the reimbursement phases. Parts of the inductor current going through C_r during the investment and the reimbursement phases are i_{ri} and i_{rr}, respectively. These parts are greyed in Fig. 5.15. The inductor current is also magnified in Fig. 5.13 and could be compared with Fig. 5.15. The dots on Fig. 5.15 show the time lapse between i_{ri} and i_{rr} that is evident in Fig. 5.13.

Considering Fig. 5.15, the reimbursement phase starts at $t = t_r$. At this moment, S_2 turns on and the inductor current increases from zero. This phase ends at $t = t_M$ when the inductor current reaches its maximal value. This part of the inductor current goes through C_V and according to Figs. 5.13 and 5.14:

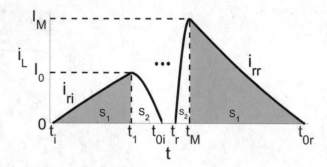

Fig. 5.15 The currents that go through C_r in the investment and the reimbursement phases in the circuit that is depicted in Fig. 5.12

$$nV_i = nV_r - \frac{1}{C_{min}} \int_{t_r}^{t_M} i_L dt. \tag{5.37}$$

The above integral can be calculated based on (5.7) and written based on k_m as follows:

$$\int_{t_r}^{t_M} i_L dt = C_{min}(nV_r - V_S)\left(1 - \frac{2}{k_m}e^{\frac{-1}{\sqrt{k_m^2-1}}cos^{-1}\left(\frac{1}{k_m}\right)}\right), \tag{5.38}$$

where V_S is the overdrive voltage of S_1 and S_2. Using (5.37) and (5.38), V_i is obtained as follows:

$$V_i = \frac{V_S}{n} + \frac{2}{k_m}\left(V_r - \frac{V_S}{n}\right)e^{\frac{-1}{\sqrt{k_m^2-1}}cos^{-1}\left(\frac{1}{k_m}\right)}. \tag{5.39}$$

Considering Fig. 5.15, the investment phase starts at t_i, when S_1 turns on and the inductor current increases from zero. This is the energizing phase that increases the stored energy in the inductor. This phase ends at $t = t_1$ when the inductor current reaches I_0. This moment is before when the inductor current reaches its maximal value, i.e. $t_1 < t_M$ and $I_0 < I_M$. At this moment, S_1 turns off, S_2 turns on, and the de-energizing phase starts. The stored current in the inductor goes through C_V, while its capacitance is maximal, $C_V = C_{max}$. The de-energizing phase ends at $t = t_{0i}$, when the inductor current reaches zero.

During the de-energizing phase of the investment phase ($t_1 \leq t \leq t_{0i}$), the voltage across C_V increases to V_r from V_i. Therefore:

$$V_r = V_i + \frac{1}{C_{max}} \int_{t_1}^{t_{0i}} i_L dt \Rightarrow \int_{t_1}^{t_{0i}} i_L dt = C_{max}(V_r - V_i). \tag{5.40}$$

The first step to calculate E_{net} in this harvester is to calculate the value of I_0 (in Fig. 5.15) for which the voltage across C_V increases to V_r during $t_1 \leq t \leq t_{0i}$. The parameter k_M is defined as following for this purpose:

$$\frac{k_M}{k_m} = \sqrt{\frac{C_{min}}{C_{max}}} \Rightarrow k_M = \sqrt{\frac{1}{n}} k_m. \tag{5.41}$$

Using (5.39) in (5.40):

$$\int_{t_1}^{t_{0i}} i_L dt = C_{max} \left(V_r - \frac{V_S}{n} \right) \left(1 - \frac{2}{k_M} e^{\frac{-1}{\sqrt{k_M^2 - 1}} \cos^{-1} \left(\frac{1}{k_M} \right)} \right). \tag{5.42}$$

The same as (5.23), the integral of i_L in the de-energizing phase of the investment phase in this harvester is as follows, where k_r is replaced with k_M and in case $k_M > 1$:

$$\int_{t_1}^{t_{0i}} i_L dt =$$

$$\frac{C_{max}}{2} \left(\sqrt{k_M^2 - 1} \; e^{\frac{1}{\sqrt{k_M^2 - 1}} \cos^{-1} \left(\frac{a_2}{\sqrt{a_1^2 + a_2^2}} \right)} \sqrt{a_1^2 + a_2^2} + a_1 - a_2 \sqrt{k_M^2 - 1} \right). \tag{5.43}$$

Since $I_0 < I_M$:

$$I_0 = g I_M, \quad g < 1. \tag{5.44}$$

I_M in the above expression is the same as in (5.20) where $V_{L1} = V_{ri} - V_S$ and k_t is replaced with k_r for this harvester. Therefore, $f(k_r)$ in the following expression is obtained from (5.20) where k_t is replaced with k_r. For the case that $k_M > 1$:

$$\begin{cases} a_1 = g \dfrac{2(V_{ri} - V_S)}{k_r} f(k_r) \\[4mm] a_2 = \dfrac{2}{\sqrt{k_M^2 - 1}} \left(g \dfrac{V_{ri} - V_S}{k_r} f(k_r) + V_i + V_S \right). \end{cases} \tag{5.45}$$

To find a closed-form expression for E_{net} for this harvester, $V_i + V_S$ is neglected in the above expression. V_i is zero for large inductance values, as the voltage across C_V changes from $n V_r$ to zero during the reimbursement phase. V_i is almost zero for smaller inductance values, since the conduction losses increase during the reimbursement phase. Therefore:

$$a_2 \approx g \frac{2(V_{ri} - V_S)}{k_r \sqrt{k_M^2 - 1}} f(k_r) \Rightarrow a_1 = a_2 \sqrt{k_M^2 - 1}. \tag{5.46}$$

With the above assumption, the expression in (5.43) is re-written as follows:

$$\int_{t_1}^{t_{0i}} i_L dt = (V_{ri} - V_S)gC_{max}\frac{k_M}{k_r}e^{\frac{-1}{\sqrt{k_M^2-1}}cos^{-1}\left(\frac{1}{k_M}\right)}.$$ (5.47)

Equating (5.42) and (5.47), g is obtained as follows:

$$g = \frac{k_r}{k_M} \cdot \frac{\left(1 - \frac{\sigma_S}{n}\right)\left(1 - \frac{2}{k_M}e^{\frac{-1}{\sqrt{k_M^2-1}}cos^{-1}\left(\frac{1}{k_M}\right)}\right)}{(\sigma_i - \sigma_S)e^{\frac{-1}{\sqrt{k_M^2-1}}cos^{-1}\left(\frac{1}{k_M}\right)}},$$ (5.48)

where:

$$\sigma_i = \frac{V_{ri}}{V_r}, \quad \sigma_S = \frac{V_S}{V_r}.$$ (5.49)

I_0 is obtained using (5.44). The expression in (5.7), where $V_{L1} = V_{ri} - V_S$, is used to express i_{ri} in Fig. 5.15. To calculate the integral of i_{ri}, the moment that $i_{ri} = I_0$ should be found. In practice, $C_r \gg C_{max}$ and therefore $t_1 \ll t_M$, where t_M is the moment that the inductor current reaches its maximal value. Therefore, i_{ri} may be approximated by a second-order function as follows, whether $k_r < 1$ or $k_r > 1$:

$$i_{ri}(t) = \frac{V_{ri} - V_S}{L}\left(t - \frac{R_s}{2L}t^2\right), \quad t_i < t < t_1.$$ (5.50)

Using the above expression and the value of I_0:

$$t_1 = \frac{1 - \sqrt{1 - \frac{2R_s I_0}{V_{ri} - V_S}}}{\frac{R_s}{L}},$$ (5.51)

therefore:

$$\int i_{ri}dt = \int_{t_i}^{t_1} i_L dt = \frac{C_r V_r (\sigma_i - \sigma_S)k_r^2}{4}\left(1 - \sqrt{1 - a_1}\right)^2\left(\frac{1}{2} - \frac{1 - \sqrt{1 - a_1}}{6}\right),$$ (5.52)

where:

$$a_1 = \frac{4}{k_r}g.$$ (5.53)

Based on the above expressions, the integral of i_{ri} is re-written as follows:

$$\int i_{ri}dt = C_r V_r y_i (k_m, n, \gamma, \sigma_i, \sigma_S). \tag{5.54}$$

where y_i can be found by comparing the above expression with (5.52).

The next step is to calculate the integral of i_{rr}, as depicted in Fig. 5.15. The expressions for the integral of i_{rr} are the same as in (5.23), (5.24), and (5.25), where V_{L1} and V_{L2} in this harvester are as follows:

$$V_{L1} = (n - \sigma_S)V_r,$$
$$V_{L2} = -(\sigma_i + \sigma_S)V_r. \tag{5.55}$$

Therefore, the integral of i_{rr} is re-written as follows:

$$\int i_{rr}dt = C_r V_r y_r (k_m, n, \gamma, \sigma_i, \sigma_S). \tag{5.56}$$

Using (2.28) and the above expressions, the net generated energy in this harvester is found:

$$E_{net} = \frac{1}{2}C_r V_r{}^2 (y_r - y_i)(2\sigma_i + y_r - y_i), \tag{5.57}$$

where y_r and y_i are the functions of the coefficients k_m, n, γ, σ_i, and σ_S. These parameters are found in the above expressions.

5.2.2.3   The Conduction Losses

The deliverable energy in this harvester is obtained by finding the enclosed area in Fig. 5.14:

$$E_{del} = \frac{1}{2}C_{max}(n-1)\left(V_r{}^2 - V_i{}^2\right), \tag{5.58}$$

where V_i is expressed in (5.39) and V_i may be expressed as a function of the following parameters:

$$V_i = y_\sigma (k_m, n, \sigma_S)V_r, \tag{5.59}$$

where y_σ can be obtained by comparing the above expression with (5.39), and therefore, E_{del} is re-written:

$$E_{del} = \frac{1}{2}C_{max}V_r{}^2(n-1)(1 - y_\sigma{}^2). \tag{5.60}$$

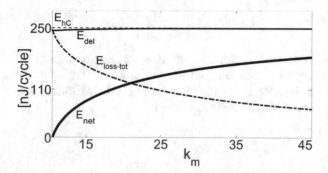

Fig. 5.16 E_{del}, E_{net}, and $E_{loss-tot}$ for the C:C-S:S harvester

Knowing E_{del} and E_{net} in (5.60) and (5.57), the total conduction losses are calculated for this harvester using (2.29).

5.2.2.4   Energy Plots

Figure 5.16 shows the deliverable energy (E_{del}), the net generated energy (E_{net}), and the total conduction losses in the C:C-S:S harvester. In this figure, $C_{min} = 1$ nF, $n = 5$, $C_r \gg C_{max}$, and $V_{ri} = V_r = 5\ V \gg V_S$, and the energies are plotted against k_m. For the selected parameters, $\gamma \approx \sigma_S \approx 0$ and $\sigma_i = 1$. In this figure, E_{hC} shows the harvested energy, and the deliverable energy is very close to this value for lower values of k_m. For higher values of k_m, $E_{del} = E_{hC}$. All the deliverable energy is converted to conduction losses for lower values of k_m. Eventually, the total conduction losses decrease and the net generated energy increases. Since $\sigma_S \approx 0$ in this figure, the total conduction reaches zero for high inductance values and $E_{net} = E_{del} = E_{hC}$.

For the selected parameters in above, E_{net}, E_{del}, and $E_{loss-tot}$ depend on C_r, C_{min}, V_r, n, and k_m. Figure 5.17 shows the impact of k_m and n on the net generated energy (E_{net}) in this harvester, where $C_r \gg C_{max}$, and $V_r = 5\ V$. Since there is a limit for the maximal feasible capacitance in a specified volume, C_{max} is fixed and equal to 5 nF for all of the plots in this figure. E_{net} does not change much after a point on each plot; therefore, it is important not to choose larger inductance values for negligible gain in E_{net}. To this end on each plot, the value of k_m is specified for which the generated energy is 90% of E_{del}. The purpose of this figure is illustrative, and these values can be found from the comprehensive expressions in this section for any other set of component values.

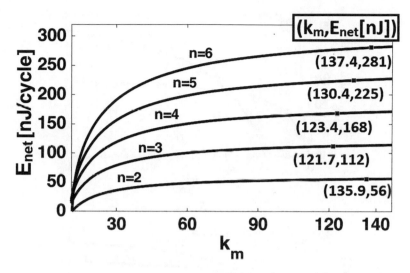

Fig. 5.17 The net generated energy in the C:C-S:S harvester versus k_m for different values of n

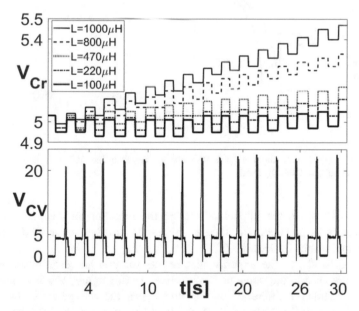

Fig. 5.18 The voltages across the reservoir capacitor and the variable capacitor in the C:C-S:S harvester for different inductance values based on the experimental results

5.2.2.5   Experimental Results

Details of an experimental setup for evaluating electrostatic harvesters are explained in Sect. 5.4. The C:C-S:S harvester in this section is tested using this setup.

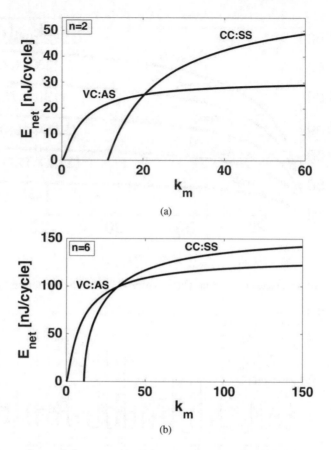

(a)

(b)

Fig. 5.19 Comparison of the net generated energy between the V:C-A:S and the C:C-S:S harvesters for (**a**) n = 2 and (**b**) n = 6

The results illustrate the impact of scaling the inductor on the performance of this harvester in practice. The theoretical results are then compared with these experimental results in Sect. 5.3.6.2.

Figure 5.18 shows the experimental results for this harvester when different inductors are used in the range of 10–1000 μH. In this figure, V_{Cr} is plotted based on the test results for different inductors and V_{CV} is the voltage across the variable capacitor in one of these experiments. As can be seen, more energy is stored in C_r when larger inductors are used for this harvester. The variable capacitor is charged to 4.4 V at the beginning of each cycle, and this is achieved by adjusting the pulse width of S_1.

5.2.2.6 Comparison of C:C-S:S and V:C-A:S Harvesters

Compared to the V:C-A:S harvester in Sect. 5.2.1, this harvester is capable of converting all the harvested energy to the net generated energy for high inductance values. However, the net generated energy in this harvester reaches zero at $k_m =$ 10.5, where E_{net} in the V:C-A:S reaches zero at $k_m = 0$. This result is obtained for the specific component values in above. Therefore, the V:C-A:S harvester performs better compared to this harvester when miniature inductors are used. E_{net} for these harvesters are plotted for $n = 2$ and $n = 6$ in Fig. 5.19. As can be seen, the performance of V:C-A:S harvester becomes superior compared to the C:C-S:S harvester for lower inductance values (lower values of k_m).

5.3 A C:C-S:S Electrostatic Harvester with a Miniature Inductor

In this section, an electrostatic harvester that operates optimally with a miniature inductor is detailed. This harvester operates under the charge-constraint scheme during both recovery and harvesting phases and has synchronous switching events to start the investment and the reimbursement phases.

5.3.1 The Circuit, Signals, QV Diagram

Figure 5.20 shows the C:C-S:S harvester with a miniature inductor that is proposed in [3]. The reservoir capacitor is split to C_{r1} and C_{r2} with the capacitance equal to $C_r/2$. Both of these capacitors are initially charged to V_{ri}. Compared to the harvesters in Sect. 5.2, V_{ri} is not fixed for all k_m values and it depends on k_m. This

Fig. 5.20 The C:C-S:S electrostatic harvester with miniaturized inductor presented in [3]

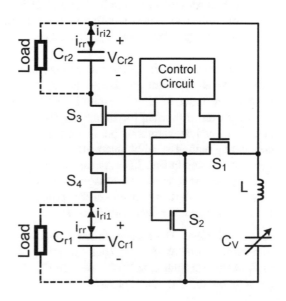

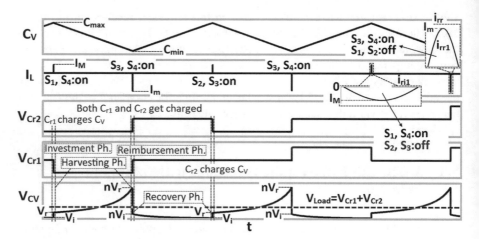

Fig. 5.21 Phases of operation, the voltages across capacitors, and the current passing through C_{r1} and C_{r2} in the C:C-S:S harvester with miniaturized inductor

dependency is discussed later in this section. The phases of operation, the voltages across C_V, C_{r1}, and C_{r2}, and the current in L for this harvester is shown in Fig. 5.21. The QV diagram for this harvester is the same as the QV diagram for the C:C-S:S harvester in Sect. 5.2. The QV diagram is depicted in Fig. 5.14. Following Figs. 5.14 and 5.21, the operation of this harvester is explained here below.

Harvesting Phase: Point **b** in Fig. 5.14 shows the start of this phase, when the capacitance of C_V is maximal and C_V is charged to V_r. During this phase, C_V is isolated (S_1–S_4: off) and the voltage across it increases from V_r to nV_r the same as explained for the harvesters in Sect. 5.2. Point **c** shows the end of this phase. In Fig. 5.21, V_{CV} shows the voltage across the variable capacitor. The harvested energy during this phase is the same as E_{hC} in (2.6).

Reimbursement Phase: When the voltage across the variable capacitor is equal to nV_r, S_3 and S_4 turn on. This connects the variable capacitor to C_{r1} and C_{r2} through the inductor which results in increasing the inductor current from zero. S_3 and S_4 turn off when the inductor current reaches zero again. Therefore, both energizing and de-energizing phases of the inductor current occur through the same conducting path. The reimbursement current, i_{rr}, that passes through both C_{r1} and C_{r2} is magnified in Fig. 5.21. As can be seen in this figure, the voltage across C_{r1} and C_{r2} increases, while the voltage across C_V decreases. In Fig. 5.14, point **c** and point **d** are the start and the end of this phase. During this phase, the voltage across the variable capacitor decreases from nV_r to nV_i (a voltage lower than nV_r).

Recovery Phase: The energy source changes the capacitance of C_V from minimal to maximal while S_1–S_4: off. Therefore, the voltage across C_V decreases from nV_i to V_i during this phase, as can be seen in Fig. 5.21. Point **d** and **a** in Fig. 5.14 mark the start and the end of this phase

Investment Phase: During the investment phase, either C_{r1} (through S_1 and S_4) or C_{r2} (through S_2 and S_3) charges the variable capacitor when $C_V = C_{max}$. Point **a** in Fig. 5.14 marks the start of this phase. Given that C_{r1} charges C_V for the first operating cycle, C_{r2} charges C_V in the second operating cycle and this goes on alternately as shown in Fig. 5.21. The inductor current increases from zero when either of C_{r1} or C_{r2} connects to C_V, and this phase ends when the inductor current reaches zero again. The same as for i_{rr}, the energizing and de-energizing phases of the inductor current occur through the same conducting path during this phase. The investment current for C_{r1} (i_{ri1}) is magnified in Fig. 5.21, and a similar investment current passes through C_{r2} (i_{ri2}) in alternating phases. The voltage across C_V at the beginning of this phase is V_i and is charged to V_r at the end of this phase: Point **b** in Fig. 5.14 is the end of this phase.

5.3.2 The Net Generated Energy

Figure 5.22 shows the currents that go through C_{r1} and C_{r2} during the investment (i_{ri}) and the reimbursement (i_{rr}) phases. The investment current is either i_{ri1} that goes through C_{r1} and C_V or i_{ri2} that goes through C_{r2} and C_V. The on and off states of the transistors for each of the cases are specified in Fig. 5.22. i_{rr} and i_{ri1} are magnified on Fig. 5.6. The dots on this figure show the time lapse between i_{ri} and i_{rr} that is evident in Fig. 5.21.

During the reimbursement phase, the voltage across C_V changes from nV_r to nV_i, while i_{rr} passes through this capacitor, therefore:

$$nV_i = nV_r - \frac{1}{C_{min}} \int i_{rr}dt \Rightarrow \int i_{rr}dt = C_{min}(nV_r - nV_i). \qquad (5.61)$$

where according to Fig. 5.22:

$$\int i_{rr}dt = \int_{t_r}^{t_{0r}} i_r dt. \qquad (5.62)$$

The voltage across C_V changes from V_i to V_r during the investment phase, while i_{ri} passes through this capacitor, hence:

$$V_r = V_i + \frac{1}{C_{max}} \int i_{ri}dt \Rightarrow \int i_{ri}dt = C_{max}(V_r - V_i), \qquad (5.63)$$

where based on Fig. 5.22:

$$\int i_{ri}dt = \int_{t_i}^{t_{0i}} i_r dt. \qquad (5.64)$$

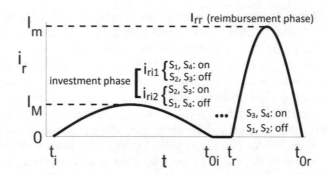

Fig. 5.22 The currents that go through C_{r1} and C_{r2} during the investment and the reimbursement phases in the circuit that is depicted in Fig. 5.20

The integral of i_{rr} is equal to the integral of i_{ri} in each operating cycle, based on (5.61) and (5.63):

$$\int i_{rr}dt = \int i_{ri}dt. \tag{5.65}$$

Therefore, given that C_{r1} has charged C_V in one operating cycle, the net stored energy (E_{net}) in C_{r1} is zero in that cycle. However, the net stored energy in C_{r2} is positive as the same reimbursement current passes through C_{r2}. Similarly, in the next operating cycle that C_{r2} charges C_V, E_{net} in C_{r2} is zero and in C_{r1} is positive. This proves that the net generated energy in each of the reservoir capacitors, C_{r1} and C_{r2}, is always positive for any value of these capacitors. In case $C_{r1} \neq C_{r2}$, E_{net} is positive but different in C_{r1} and C_{r2}, and no matching between these capacitors is required. Hence, the following energy is stored in either C_{r1} or C_{r2} in each operating cycle, according to (2.28):

$$E_{net} = \frac{1}{2}\int i_{rr}dt\left(2V_{ri} + \frac{1}{C_{r1}}\int i_{rr}dt\right). \tag{5.66}$$

In the above expression, the integral of i_{rr} and V_{ri} are the parameters that should be found to calculate E_{net} for this harvester. In obtaining these parameters as follows, it is assumed that $C_{r1} = C_{r2} = C_r/2$. The expression for i_{rr} is the same as in (5.7)–(5.8), where:

$$V_{L1} = nV_r - 2V_{ri} - 2V_S, \quad C_{t1} = C_{tm}, \tag{5.67}$$

where C_{tm} is the series equivalent of C_{r1}, C_{r2}, and C_{min}. Therefore:

$$\int i_{rr}dt = \alpha C_{min}(nV_r - 2V_{ri} - 2V_S)(1 + g_2), \tag{5.68}$$

where:

$$g_2 = \begin{cases} e^{-\frac{\pi}{\sqrt{\frac{k_m^2}{\alpha} - 1}}} & k_m > \sqrt{\alpha} \\ 0 & k_m < \sqrt{\alpha} \end{cases} \tag{5.69}$$

and:

$$\alpha = \frac{1}{1 + 4\frac{C_{min}}{C_r}} = \frac{1}{1 + 4\gamma}. \tag{5.70}$$

During the investment phase, C_{r1} (or C_{r2}) connects to the variable capacitor and i_{ri} goes through C_{r1} (or C_{r2}) and C_V. At the end of this phase, C_V should be charged to V_r. The switching time may not be controlled to charge C_V to V_r, since C_{r1} (or C_{r2}) and C_V are connected until i_{ri} reaches zero. Therefore, V_{ri} (the initial voltage across C_{r1} or C_{r2}) should be adjusted in a way that C_V is charged to V_r at the end of this phase. To find the adjusted V_{ri}, the following steps are taken.

The expression for i_{ri} is the same as in (5.7)–(5.8), where:

$$V_{L1} = V_{ri} - V_i - 2V_S, \quad C_{t1} = C_{tM}, \tag{5.71}$$

where C_{tM} is the series equivalent of C_{max} and C_{r1} (or C_{r2}). Therefore:

$$\int i_{ri} dt = n\beta C_{min}(V_{ri} - V_i - 2V_S)(1 + g_1), \tag{5.72}$$

where:

$$g_1 = \begin{cases} e^{-\frac{\pi}{\sqrt{\frac{k_m^2}{n\beta} - 1}}} & k_m > \sqrt{n\beta} \\ 0 & k_m < \sqrt{n\beta} \end{cases} \tag{5.73}$$

and:

$$\beta = \frac{1}{1 + 2n\frac{C_{min}}{C_r}} = \frac{1}{1 + 2n\gamma}. \tag{5.74}$$

It is shown that the integral of i_{rr} and the integral of i_{ri} are equal in (5.65). Equating the integral of i_{rr} as in (5.68) and the integral of i_{ri} as in (5.72) and using (5.61), V_{ri} is obtained:

$$V_{ri} = \eta V_r,$$

$$\eta = \frac{\alpha(n - 2\sigma_S)(1 + g_2) + n\beta(1 + 2\sigma_S)(1 + g_1) - \alpha\beta(n - 2\sigma_S)(1 + g_1)(1 + g_2)}{2\alpha(1 + g_2) + n\beta(1 + g_1) - 2\alpha\beta(1 + g_1)(1 + g_2)}, \quad (5.75)$$

where σ_S is defined in (5.49). In the above expressions, the acceptable range for η is $0 < \eta \leq 1$. It can be proved that for any value of γ and k_m, η would be out of this acceptable range if $1 < n < 2$. Therefore, this harvester needs a variable capacitor with $n > 2$ to operate.

Using (5.63) and (5.72), V_i is found:

$$V_i = \left(1 - \alpha\left(1 - \frac{2}{n}\eta\right)(1 + g_2)\right)V_r. \quad (5.76)$$

Considering the expressions for the integral of i_{rr} and V_{ri} in (5.68) and (5.75), and replacing them in (5.66), E_{net} is expressed as follows:

$$E_{net} = \alpha C_{min} V_r^2(n - 2\eta - 2\sigma_S)\left(\eta + \frac{1 - \alpha}{4}(n - 2\eta - 2\sigma_S)(1 + g_2)\right)(1 + g_2). \quad (5.77)$$

5.3.3   The Conduction Losses

The QV diagram for this harvester is the same as the C:C-S:S harvester in Sect. 5.2.2, depicted in Fig. 5.14. Therefore, the expression for E_{del} of this harvester is the same as in (5.58), where V_i for this harvester is expressed in (5.76). V_i in (5.76) is a function of the parameters, k_m, n, γ, and σ_S as follows:

$$V_i = y(k_m, n, \gamma, \sigma_S)V_r, \quad (5.78)$$

therefore, E_{del} for this harvester is

$$E_{del} = \frac{1}{2}C_{max}(n - 1)(1 - y^2). \quad (5.79)$$

Knowing E_{net} and E_{del} in (5.77) and (5.79), the total conduction losses are calculated for this harvester using (2.29). According to aforementioned expressions, E_{net}, E_{del}, and $E_{loss-tot}$ for this harvester can be expressed as functions of k_m, n, γ, and σ_S.

5.3.4   Energy Plots

Figure 5.23 shows the deliverable energy (E_{del}), the net generated energy (E_{net}), and the total conduction losses in the C:C-S:S harvester with a miniature inductor.

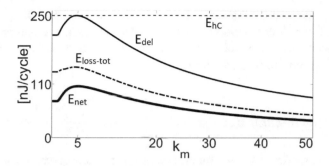

Fig. 5.23 E_{del}, E_{net}, and $E_{loss-tot}$ for the C:C-S:S harvester with miniaturized inductor

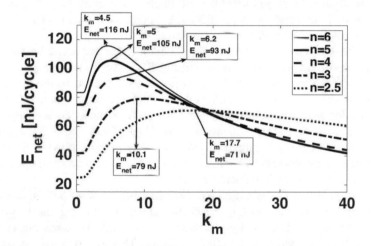

Fig. 5.24 The net generated energy in the C:C-S:S harvester with miniaturized inductor versus k_m for different values of n

In this figure, $C_{min} = 1\,\text{nF}$, $n = 5$, $C_r \gg C_{max}$, and $V_{ri} \gg V_S$ ($\sigma_S = 0$), and the energies are plotted against k_m. Compared to the conventional harvesters in Sect. 5.2, the net generated energy in this harvester is not zero without any inductor, $k_m = 0$. For the selected component values, this harvester generates 75 nJ/cycle net energy when no inductor is present in the harvester circuit. For extremely low inductance values ($k_m < 1$), all the energies are fixed and equal to when no inductor is used. Eventually, E_{del}, $E_{loss-tot}$, and E_{net} increase for higher inductance values. At a low value of k_m ($k_m = 5$) the deliverable energy (E_{del}) equals to the harvested energy (E_{hC}). At this point, the net generated energy of the harvester peaks. The net generated energy of this harvester starts decreasing after this point. Therefore, this harvester operates optimally with a miniaturized inductor.

For the selected parameters in above, E_{net}, E_{del}, and $E_{loss-tot}$ depend on C_r, C_{min}, V_r, n, and k_m. Figure 5.24 shows the impact of k_m and n on the net generated energy (E_{net}) in this harvester, where $C_r \gg C_{max}$, and $V_r = 5\,\text{V}$. Since there is a limit for the maximal feasible capacitance in a specified volume, C_{max} is fixed and equal to 5 nF for all of the plots in this figure. On each plot, the optimal operating point is specified. At this point, the harvester generates maximal energy for the selected values for C_r, C_{min}, n, and V_r.

5.3.5   Experimental Results

Details of an experimental setup for evaluating electrostatic harvesters are explained in Sect. 5.4. The C:C-S:S harvester with a miniature inductor in this section is tested using this setup. The results illustrate the impact of scaling the inductor on the performance of this harvester in practice. The theoretical results are then compared with these experimental results in Sect. 5.3.6.

The voltage across C_{r1} is plotted for different inductors based on the experimental results of the proposed harvester in Fig. 5.25. The initial voltage across this capacitor and C_{r2} for each inductor is adjusted so that each of these capacitors charge the variable capacitor to 4.4 V at the beginning of each cycle. Therefore, in all the tests for this harvester, the harvested energy from the external energy source is the same as in all the experiments for the V:C-A:S and C:C-S:S harvesters in Sect. 5.2. The increase of the stored energy in C_{r1} (ΔE_{Cr1}) is plotted to visualize the performance of this harvester for different inductors. As can be seen, the harvester operates best when a 47 µH inductor is used.

The voltage across C_{r1} is monitored through a buffer. Nonetheless, the voltage increases across both C_{r1} and C_{r2} reflect the net generated energy in the proposed harvester. Therefore, $V_{Cr1} + V_{Cr2}$ is measured through the same buffer before the harvester starts operating and after it finishes. This measurement procedure is plotted in Fig. 5.25 for $L = 47$ µH. To ensure that no charge is injected into the core harvester through the monitoring buffers, the experiments were repeated without these buffers, and identical results were obtained.

5.3.6   Comparison with Conventional Harvesters

In calculating the energies for any harvester in this book, it is assumed that C_V is charged to the same parametric voltage, V_r at the beginning of the harvesting phase. To achieve this goal in this harvester, the value of V_{ri} should be adjusted for any k_m according to (5.75). This way, the voltage across the variable capacitor changes from V_r to nV_r during the harvesting phase for any value of inductance. Therefore, selecting a larger inductor does not change the amount of energy that the variable capacitor harvests from the external energy source. Subsequently, the efficiencies of the harvesters are compared when they all harvest the same amount of energy from the external energy source.

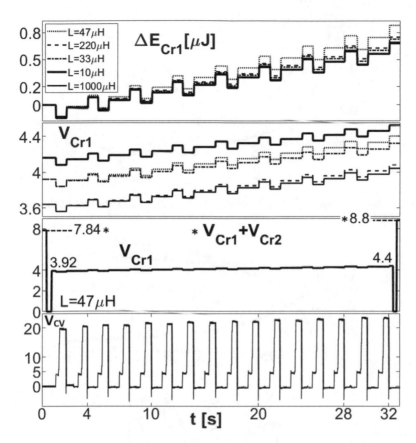

Fig. 5.25 The voltages across the reservoir capacitors and the variable capacitor in the C:C-S:S harvester with miniaturized inductor for different inductance values based on the experimental results

The net generated energy for the C:C-S:S harvester with a miniature inductor depends on V_{ri} and $\int i_{rr}$ based on (5.66). Figure 5.26 shows the normalized values of these two parameters against k_m for $C_{min} = 1\,nF$ and $n = 5$ where $C_r \gg C_{max}$. As can be seen, V_{ri} decreases for the higher values of k_m: lower V_{ri} across C_{r1} (and C_{r2}) is required to charge C_V to V_r during the investment phase, when the quality factor of the investment path improves. As expected, the normalized integral of i_{rr} increases for higher values of k_m, not only because the quality factor of the conducting path in the reimbursement phase improves but also because V_{ri} decreases (based on (5.68)). As can be seen in Fig. 5.24, the optimized k_m for this harvester is $k_m = 5$ for the aforementioned values of C_{min} and n, due to the opposite effect of increasing k_m on V_{ri} and $\int i_{rr}$.

The initial voltage across C_r, V_{ri}, is constant for any value of k_m for conventional harvesters (V:C-A:S and C:C-S:S harvester) in Sect. 5.2. The integral of i_{rr} in the V:C-A:S harvester and the difference between the integral of i_{rr} and the integral of

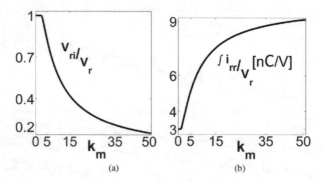

(a)   (b)

Fig. 5.26 (a) Normalized V_{ri} and (b) normalized $\int i_{rr}$ versus k_m for the C:C-S:S harvester with miniaturized inductor

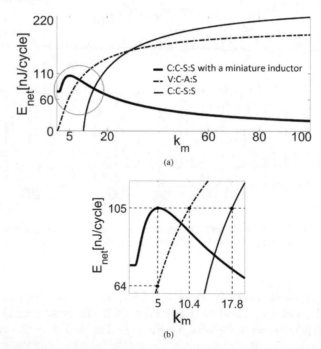

(a)

(b)

Fig. 5.27 (a) Zoomed out and (b) zoomed in (for lower inductance values) comparison of E_{net} for the conventional V:C-A:S, C:C-S:S, and C:C-S:S with a miniature inductor harvesters

i_{ri} in the C:C-S:S harvester increase as k_m increases. These integrals increase for higher values of k_m as the quality factor of conducting paths improves at higher inductance values. Therefore, the net generated energy in these harvesters increases for higher values of k_m, as depicted in Figs. 5.9 and 5.16.

5.3.6.1   Theoretical Comparison

Figure 5.27a shows the comparison between the harvester in this section and the conventional harvesters in Sect. 5.2 versus variations in k_m. These plots are drawn for the case that $C_r \gg C_{max}$ and $n = 5$. As can be seen, the conventional C:C-S:S harvester is capable of storing the most possible energy in its reservoir capacitor when a large inductor is used. However, this harvester performs the worst for low inductance values. It can be observed that the C:C-S:S harvester with a miniature inductor performs the best for low inductances compared to the other two harvesters.

The region of interest in Fig. 5.27a is magnified in Fig. 5.27b. The vertical axis shows the amount of stored energy in each operating cycle. As can be seen in this figure, there is an optimal k_m for which the C:C-S:S harvester with a miniature harvester stores the most energy in its reservoir capacitors. At this point, the harvester stores 1.64 times more energy in its reservoirs capacitors compared to conventional V:C-A:S harvester. However, the conventional C:C-S:S harvester is incapable of storing any energy with this inductance. Equivalently, the conventional V:C-A:S needs a 4.5 times larger inductor and the conventional C:C-S:S needs a 13.1 times larger inductor to store the same amount of energy as the optimal net generated energy in the C:C-S:S with a miniature inductor.

5.3.6.2   Experimental Comparison

In Fig. 5.28, the experimental results for inductors ranging from 1 to 1000 μH are shown with dashed lines while the solid lines are based on the derived expressions. As can be seen, the experimental results for all of the harvesters are very close to the expected plots.

The optimal operating point of the C:C-S:S harvester with a miniature inductor is obtained when $k_m = 3.36$ based on the experimental results in Fig. 5.28. At this

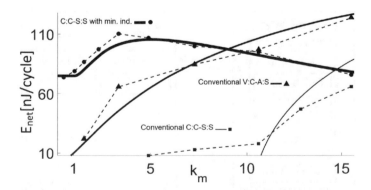

Fig. 5.28 Comparison of E_{net} between the harvester in this section and the conventional harvesters in Sect. 5.2 for lower inductance values, experimental: - -, derived expressions: –

point, the harvester uses a 47 μH inductor and generates 110.3 nJ energy in each operating cycle. Using this inductor, the conventional C:C-S:S harvester does not generate a positive energy and the conventional V:C-A:S generates 65.8 nJ energy in each operating cycle. Moreover, V:C-A:S harvester should employ a 735 μH inductor to generate around 110 nJ energy in each operating cycle. This inductor is 15.6 times larger than the inductor (47 μH) that the C:C-S:S harvester with a miniature inductor needs for generating this much energy. The conventional C:C-S:S harvester generates only 66 nJ in each operating cycle employing a 1000 $u\mu$H inductor. The experimental results validate the high performance of the C:C-S:S harvester with a miniature inductor, when smaller inductance values should be used.

5.4 Experimental Setup

This section discusses the practical considerations in setting up an experimental test bench to evaluate the performance of electrostatic harvesters. Using this test bench, the performance of the harvesters in Sects. 5.2 and 5.3 is investigated in Sect. 5.3.6. The development of the experimental setup in this section is explained mainly based on discrete components. However, similar considerations could be applied to an integrated experimental setup.

5.4.1 Required Components

Figure 5.29 outlines the required components for implementing switching electrostatic harvesters. These components may be categorized under two main blocks, i.e. harvester core and control block.

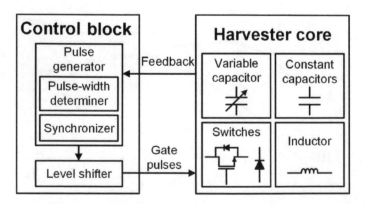

Fig. 5.29 The required control blocks in switching electrostatic harvesters

5.4.1.1 Harvester Core

In selecting each of the components in the harvester core a number of considerations should be taken into account. These considerations may be general or application specific. As an example, avoiding bulky components in applications with a limiting form-factor is an application specific consideration. Selecting transistors with low quiescent current is a general consideration. However, in applications where the energy source has a low frequency or the deliverable energy is low, more stress is put on this consideration.

Variable capacitor: Each capacitor has a self-discharge time constant that is defined as follows:

$$\tau_{dis} = R_{ins}C, \tag{5.80}$$

where C is the capacitance of the capacitor and R_{ins} is the insulation resistance of the capacitor. The insulation resistance is mainly dependant on the material specifications of the insulating layer between the plates of the capacitor. The capacitor loses about 37% its initial charge after $t = \tau_{dis}$. The capacitance of a variable capacitor changes between $C = C_{min}$ and $C = C_{max}$. As a result, C and R_{ins} and therefore τ_{dis} are not fixed for a variable capacitor. Although it depends on the mechanics of the variable capacitor, a lower τ_{dis} is expected for when the capacitance of the variable capacitor is minimal.

In practice, the time period of an energy conversion cycle should be much less than τ_{dis}. Otherwise, self-discharge of the variable capacitor impacts the performance of the harvester drastically. Therefore, care must be exercised in selecting a variable capacitor for applications where the frequency of the energy source is low. In these applications, the variable capacitor should have a high τ and/or be able to convert the low frequency of the energy source to a high frequency of changes in its capacitance. The more the frequency of changes in the capacitance of the variable capacitor, the less time period for the energy conversion cycle.

Another consideration in selecting or designing a variable capacitor for the experimental setup is eliminating causes for generating extra electrostatic charges. An example is a variable capacitor with a dielectric layer being rubbed against the rotating plates and the stationary plates. In this variable capacitor, unexpected electrostatic charges are generated that impact the voltage across the variable capacitor. This effect is not considered in the theoretical analyses presented in this book. Avoiding this structure for the variable capacitor is necessary when the experimental setup is used to verify theoretical results.

Constant capacitors: The self-discharge time constant (τ_{dis}) should be much higher for the constant capacitors compared to the variable capacitor, as they need to keep charge in themselves for the duration of the whole test. The capacitance of constant capacitors is several times bigger than the maximum capacitance of the variable capacitor, so that the storing capabilities of them are maximized.

This makes the constant capacitors beneficial in terms of self-discharge time constant compared to the variable capacitor. However, the fabrication technology of the constant capacitors has a huge impact on the insulation resistance. The insulation resistance of the Multilayer Ceramic capacitors is much higher compared to electrolytic capacitors.

Inductor: In selecting an inductor, its saturation current should be noted. This current should always be less than the actual inductor current in a circuit. The inductor current increases dramatically more than expected, when a current more than its saturation current passes through it. The inductor current during the energizing phase is shown in Fig. 5.3a. As can be seen, the maximal inductor current is higher for lower values of k_1. Therefore in electrostatic harvesters, miniaturized inductors need to have a higher saturation current compared to large inductors.

Switches: Using transistors with less quiescent current and diodes with less reverse current is necessary for minimizing the quiescent losses. In Sects. 5.2 and 5.3, higher values of σ_S and σ_F increase the conduction losses. To minimize σ_F, diodes with low forward voltage drop should be selected. Transistors should be biased in deep-triode region to minimize the conduction losses in them while they are conducting. In this region, the transistor is modelled with an on-resistance when it is conducting: The overdrive voltage is zero and $\sigma_S = 0$. This is explained in more details in Sect. 3.1.1.

5.4.1.2   Control Block

The control block in Fig. 5.29 generates the gate pulses with appropriate pulse-width and at the specific moments (when $C_V = C_{min}$ and $C_V = C_{max}$) for switching the transistors in the harvester core. Pulse-width determiner, synchronizer, and level shifter are the representative blocks for this task. The pulse-width determiner specifies delay, period, and duty cycle of these gate pulses and the synchronizer asserts the switching moment. Both of these blocks determine the parameters of the gate pulses based on the feedback signals from the harvester core. The level shifter is required to generate gate pulses for transistors with their sources connected to a higher voltage.

In [1], an ASIC is developed that implements all the control blocks mentioned in Fig. 5.29 by monitoring voltages across two added sensing capacitors and the voltage across the inductor. In [2], an IC is fabricated with a programmable delay line for adjusting the gate pulses. This IC includes power transistors of the circuit. The IC does not include a synchronization block: However, digital methods for synchronizing the control circuit and the core harvester based on energy feedback are explained and simulated.

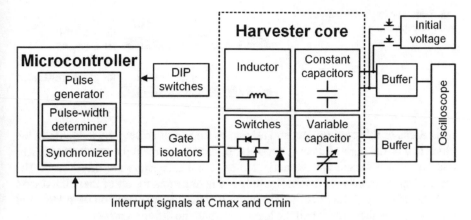

Fig. 5.30 The implemented block diagram for the electrostatic harvesters in Sects. 5.2 and 5.3

5.4.1.3 Measuring Configuration

The main parameter of interest in energy harvesting applications is the amount of energy that is stored in the constant capacitors as a result of the harvester operation. The common solution is to measure the voltage increase across these capacitors during the operation of the harvester, while no load is in parallel with them. Figure 5.30 shows the implemented experimental setup. In this setup, the voltage across the constant capacitors and the variable capacitor are monitored continuously to follow the exact behaviour of the energy harvesting circuit and to detect any failure during the operation of the harvester. The desired nodes are connected to the oscilloscope through a buffer.

The constant capacitors that are selected in the performed experiments are in the range of few hundreds of nF. The oscilloscope probes normally have a series $10 M\Omega$ resistance. This means that the charge in constant capacitors gets depleted in few seconds if the probe is connected directly to them according to $\tau = RC$. This problem is much worse for the variable capacitor which changes in the range of few nF. High gain Opamps are used as buffers between the probe and the desired node to tackle this problem, as can be seen in Fig. 5.30. The voltages across the constant capacitors and the variable capacitor are in the range of 4–8 and 20–40 V, respectively. Therefore, two different Opamps with the supply voltages of ± 18 and ± 50 V are employed for the constant capacitors and the variable capacitor.

5.4.2 Implemented Experimental Setup

To draw a fair comparison, the conventional harvesters in Sect. 5.2 and the harvester in Sect. 5.3 are implemented and operated using the same variable capacitor actuated under the same circumstances. This comparison is presented in Sect. 5.3.6. The

Fig. 5.31 The implemented
variable capacitor

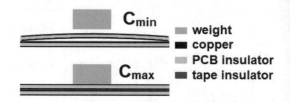

main focus of this comparison is to assess the impact of scaling the inductor on the
efficiency of the core of these harvesters. In this comparison the conduction losses
and losses due to leakage currents of diodes and transistors are considered. The
same control circuit with external supply is implemented for all of these harvesters
as depicted in Fig. 5.30. This way, the loading of the control circuit on the nodes used
for feedback is the same for all the harvesters, and the different power consumptions
of the control circuits are eliminated. In this comparison, no external power source
is connected to the core of the harvesters while operating, hence all the component
losses are included and a more accurate comparison of the core of the harvesters is
achievable.

In Fig. 5.30, a microcontroller produces gate pulses when an interrupt signal is
generated. This interrupt signal is generated when C_V is maximal and minimal.
This ensures that the operation of all the harvesters is synchronized with C_V. The
harvesters do not operate optimally if the switching events are not synchronized
with the moments that C_V is maximal and minimal. Gate isolators are used for level
shifting the pulses generated by the microcontroller. Each harvester is tested with
different inductors, and the width of the gate pulses for these inductors is adjusted
with dip switches. The reservoir capacitor(s) are initially charged through push
buttons. The voltages across the reservoir capacitor(s) and C_V are monitored with
buffer circuits to eliminate the loading effects of the oscilloscope probe on these
nodes, as explained in Sect. 5.4.1.3.

A variable capacitor is implemented to be used as the transducer for all of the
three harvesters as shown in Fig. 5.31. This guarantees that the harvested energy
from the external energy source is the same for all the harvesters, since the same
variable capacitor has been used. Two double sided PCB sheets facing each other are
insulated by an adhesive tape layer and form a variable capacitor. The copper layers
facing out protect the variable capacitor against unwanted electrostatic charges. A
weight is used to change the capacitance of the variable capacitor. The implemented
variable capacitor has a minimum capacitance of 1 nF, a maximum capacitance of
5 nF, and a parallel resistance exceeding 1000 MΩ. The high parallel resistance is
crucial due to the low operating frequency in the following tests. A weight is placed
over the variable capacitor for 1 s and is removed from over it during the next 1 s,
and this goes on alternately to imitate low frequency body movements, e.g. the
diaphragm muscle [4] and knee joint movements [5].

Figure 5.32 shows the fabricated PCBs for the three discussed harvesters.
Discrete NMOS and PMOS transistors (BSS123 and BSS84) and isolated gate
drivers are used to implement the switches of these harvesters. A back to back

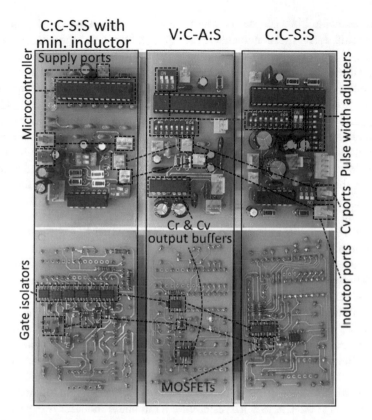

Fig. 5.32 The fabricated PCBs for conventional harvesters in Sect. 5.2 and the harvester with miniaturized inductor in Sect. 5.3

NMOS structure is employed to implement the bidirectional switches wherever required, due to the existence of intrinsic diodes in discrete transistors. Each harvester is tested with different inductors and the same variable capacitor. To this end, each PCB has an input port for the variable capacitor and an input port for the inductor. Each PCB has four ports to supply voltages of the microcontroller, gate drivers, buffer, and initial voltage across the reservoir capacitors.

5.5 Conclusion

Table 5.1 summarizes the comparison of the three harvesters including the minimum number of passive and active discrete or integrated switches required for each design. A part of the net generated energy should be used to power the control circuit which generates gate pulses for the transistors in all of these harvesters. The power consumption of this control unit depends on the number of active switches and the moments it should detect when C_V is minimal or maximal. The control circuit should detect these moments if the harvester has a synchronous event at the beginning of the investment phase and/or the reimbursement phase.

The values of C_{max} and C_{min} may vary during the operation of the harvester based on the characteristics of the variable capacitor and the kinetic energy source. The pulse width of the gate signals in the conventional V:C-A:S and the C:C-S:S harvesters is dependent on these variations. The variations in the values of C_{max} and C_{min} deteriorate the performance of these harvesters. However, the pulse width of the gate signals could be any longer than T/2 ($T = 2\pi\sqrt{LC_{max}}$) in the C:C-S:S harvester with a miniature inductor because of the intrinsic diodes that are present in the discrete MOSFETs. These diodes block the current in the opposite direction. Therefore, the net generated energy is much less sensitive to the variations of C_{max} and C_{min} in this harvester. The switching scheme for this harvester is also capable of reducing this sensitivity when implemented with integrated transistors. The minimum required n (C_{max}/C_{min}) for the conventional harvesters in Sect. 5.2 is 1. However, this parameter of the variable capacitor should be larger than 2 in the C:C-S:S harvester with a miniature inductor. Therefore, this harvester is not able to operate when a variable capacitor with n less than 2 is used.

The implementation efficiency of the control circuit for the conventional V:C-A:S and C:C-S:S harvesters is stated in Table 5.1 according to the reported values of the control circuit power consumption and the net generated energy in [1] and [2] when a large inductor is used. The net generated energy in these two harvesters decreases for smaller inductors, and this worsens the control circuit's implementation efficiency. Nonetheless, a higher implementation efficiency for the control circuit of the C:C-S:S harvester with a miniature inductor is expected when a miniature-size inductor should be used. The same control circuit as outlined in Fig. 5.30 is implemented for all of the three harvesters with discrete components and supplied with an external power source since the main focus is on the comparison between the harvester core in these structures. An integrated implementation of the control circuit for the C:C-S:S harvester with miniaturized inductor could be considered in future works.

The harvesters that are based on using an inductor often use a large inductor to eliminate the conduction losses. In this section, a variable-capacitance harvester is discussed that outperforms conventional V:C-A:S and C:C-S:S charge-constraint harvesters while using smaller inductors. All three harvesters are implemented and tested with the same variable capacitor having a minimum capacitance of 1 nF and $n = 5$. The harvesters were tested with an actuating frequency as low as 0.5 Hz to validate their ability of harvesting energy from low frequency energy sources. The experimental results show that to generate optimal harvested energy of the C:C-S:S harvester with a miniature inductor, the conventional V:C-A:S harvester has to employ a 15.6 times larger inductor. At the optimal inductance value for the C:C-S:S harvester with a miniature inductor, the conventional C:C-S:S harvester is incapable of generating positive energy. At this inductance value, the C:C-S:S harvester with a miniature inductor generates 1.67 times more energy than the conventional V:C-A:S harvester.

Table 5.1 Comparison of performance, number of switches, control circuit requirements, and its power consumption

Size of inductor	Performance of the energy harvesting circuit E_{net} [nJ/cycle]				Min. n	Number of switches			Control circuit		Implementation efficiency[a]
						Discrete		Integ.	Moments to be detected	Pulse-width sensitivity to C_V variations	
	Zero	Miniature	Medium	Large		D	MOS	MOS			
k_m	0	3.36	15.5	1000							
Evaluation solution	Tested			Calcul.							
Conventional V:C-A:S Harvester	*0*	*65.8*	*123.9*	*198.3*	1	1	2	3	C_{min}	*Yes*	2.9%
Conventional C:C-S:S Harvester	*0*	*0*	*66.1*	*246.5*	1	0	2	2	C_{min}, C_{max}	*Yes*	29%
C:C-S:S Harvester with Min. Ind.	*55.8*	*110.3*	*76.1*	*N/A*	2	0	6	4	C_{min}, C_{max}	*No*	D/I[b]

[a]Ratio of the control circuit power consumption to E_{net} as reported in [2] and [1] when a large inductor is employed

[b]Discrete Implementation with external power supply: not comparable with the integrated control units in [2] and [1]

References

1. A. Kempitiya, D.A. Borca-Tasciuc, M.M. Hella, Low-power ASIC for microwatt electrostatic energy harvesters. IEEE Trans. Ind. Electron. **60**(12), 5639–5647 (2013)
2. S. Meninger, J.O. Mur-Miranda, R. Amirtharajah, A. Chandrakasan, J.H. Lang, Vibration-to-electric energy conversion. IEEE Trans. Very Large Scale Integr. (VLSI) Syst. **9**(1), 64–76 (2001)
3. S. H. Daneshvar, M. Maymandi-Nejad, M.R. Yuce, J.-M. Redouté, A variable-capacitance energy harvester with miniaturized inductor targeting implantable devices. IEEE Trans. Ind. Electron. **69**(1), 475–484 (2022)
4. S.H. Daneshvar, M. Maymandi-Nejad, A new electro-static micro-generator for energy harvesting from diaphragm muscle. Int. J. Circuit Theory Appl. **45**(12), 2307–2328 (2017)
5. S.H. Daneshvar, M. Maymandi-Nejad, M. Sachdev, J. Redouté, A charge-depletion study of an electrostatic generator with adjustable output voltage. IEEE Sens. J. **19**(3), 1028–1039 (2019)

Conclusion

Energy harvesting from kinetic energy sources is a promising powering solution in many emerging applications including IoT, sensor networks, implantable medical devices, and new medical treatments. Unique features of electrostatic harvesters make them superior in a wide range of these applications. Fundamentals, analysis techniques, and practical insights are discussed for different categories of variable-capacitance electrostatic harvesters in this book. Each category of electrostatic harvesters has its own advantages and disadvantages in terms of required components and volume, net generated energy, conduction losses, energy efficiency, and control circuit complexity. Therefore, there is not a single structure performing the best in different applications. Having an in-depth understanding of the features of these categories, one may select the most appropriate electrostatic harvester according to the application specifications and requirements.

© The Author(s), under exclusive license to Springer Nature Switzerland AG 2022   205
S. H. Daneshvar et al., *Design of Miniaturized Variable-Capacitance Electrostatic Energy Harvesters*, https://doi.org/10.1007/978-3-030-90252-0

Printed in the United States
by Baker & Taylor Publisher Services